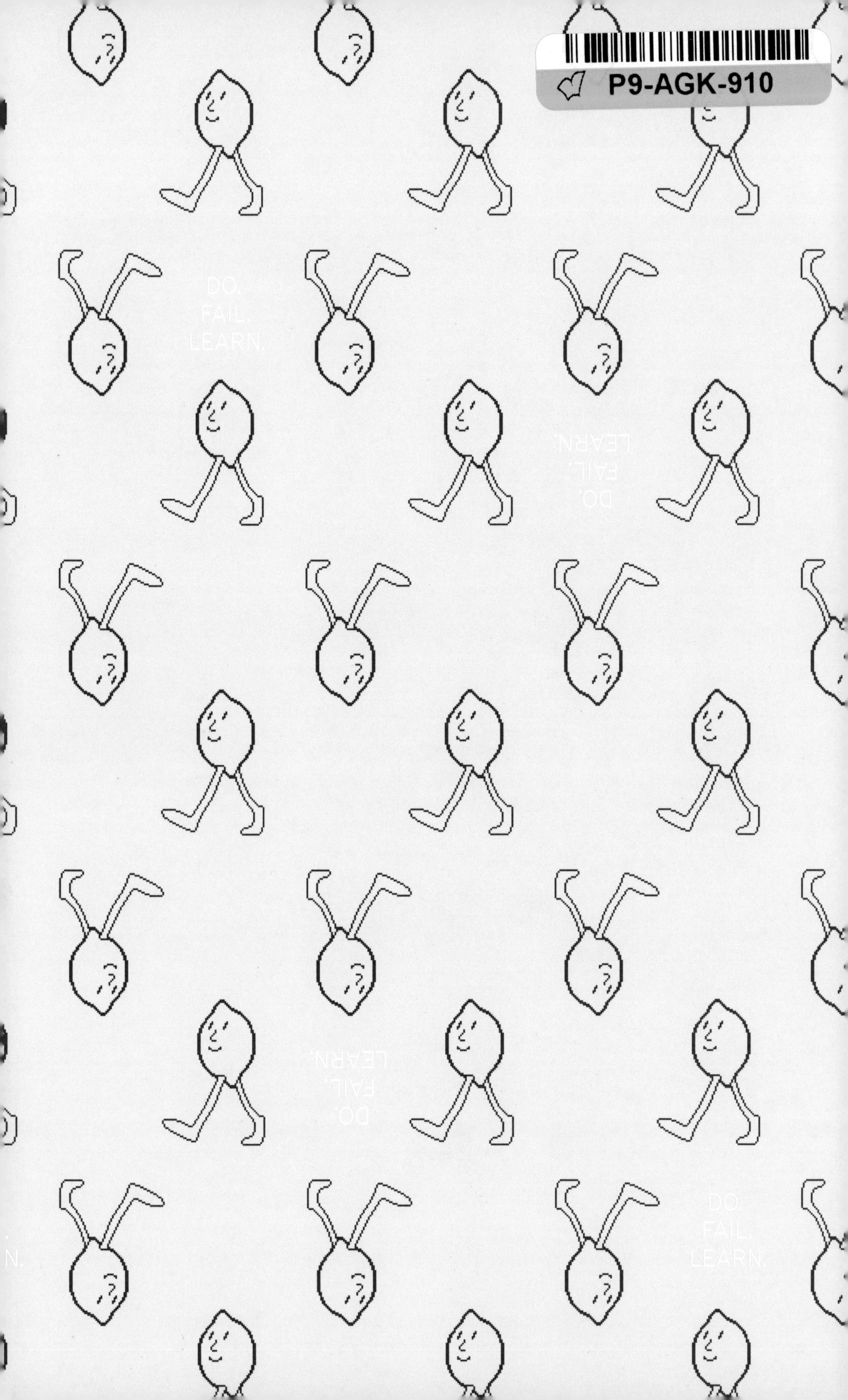
DO.
FAIL.
LEARN.
P9-AGK-910

PRAISE FOR *BUILD*

"Tony's insights, instincts, and wisdom are essential reading and precious gifts for any inventor hungry to change the world."

—Thomas Heatherwick, award-winning designer and founder, Heatherwick Studio

"Tony has an uncanny ability to take his infinite wisdom and legacy of leadership, mentorship, innovation, and success and break it down into a practical, no-nonsense how-to guide on what it means to build something with meaning that will endure."

—Daniel Ek, founder and CEO, Spotify

"You owe it to yourself to benefit from Tony's amazing journey and the advice he received from iconic mentors like Steve Jobs and Bill Campbell. Candid, often bold straight talk that will help you immensely on your journey."

—Bill Gurley, general partner, Benchmark

"Fadell leaves no stone unturned in the art and science of turning ideas into reality. This book is pure muscle for the next generation of product leaders and outfits us all with actionable tactics. Absorb it deeply and then get to work."

—Scott Belsky, founder, Behance and CPO, Adobe

"*Build* is the new bible for anyone interested in creating successful products and companies. Fadell's frank account of an epic period in Silicon Valley is so engrossing, you won't realize how much you have learned until the finish."

—Randy Komisar, general partner, Kleiner Perkins and author, *The Monk and the Riddle*

"Tony Fadell takes the reader on a rollercoaster ride, revealing what it's really like to work in Silicon Valley and simultaneously providing a compelling instruction manual for anyone who wants to follow in his footsteps."

—John Markoff, author, *Machines of Loving Grace*

"Tony's experience can be applied to any builder or creator anywhere in the world. The challenges are always the same, and Tony shares a number of insights on how to navigate them."

—Micky Malka, managing partner, Ribbit Capital

"Truly historic anecdotes, backstories, and straight-talk advice from a hall of fame entrepreneur who has seen and done it all. I thought I'd be too busy to read this book once. I read it twice. TWICE."

—Jim Lanzone, CEO of Yahoo

"This isn't your typical business book. It's a loud, passionate, mission-driven anthem about how to build, from one of the greatest product designers of all time. If you want to learn something about how to start a company, be a CEO, design world-changing products, or build anything great, this is it, right here."

—Steve Vassallo, general partner, Foundation Capital

"This is an insightful overview into Tony's extraordinary life at the junction of the biggest technological revolution of humankind. It is an amazing blueprint on how creative thinkers can negotiate their way to making ideas come to life in the world."

—Sir David Adjaye, OBE, architect

"Tony Fadell is the rare combination of engineer and entrepreneur, with the soul of a storyteller. *Build* takes you far away from the beige box and into the glistening white world of the man who gave us the iPod, iPhone, and Nest, while providing actionable advice for formulating, launching, scaling, and even selling a business."

—Benjamin Clymer, founder, HODINKEE

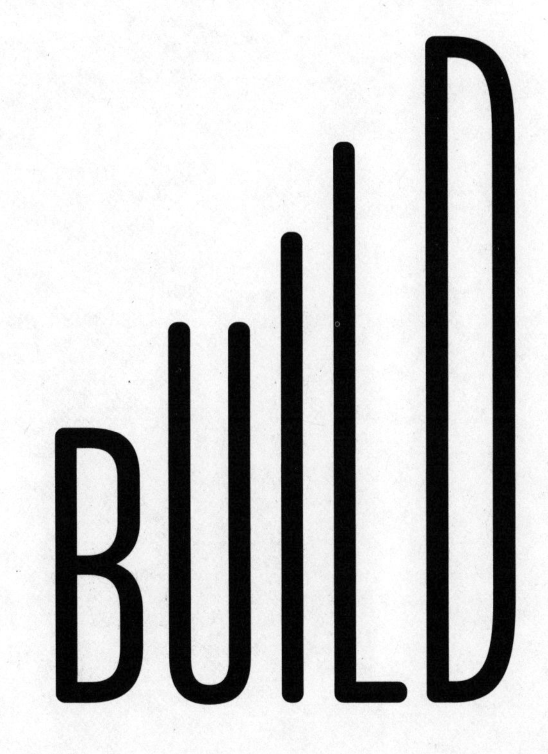
BUILD

BUILD

An Unorthodox Guide to Making Things Worth Making

TONY FADELL

HARPER BUSINESS
An Imprint of HarperCollins*Publishers*

HarperCollins books may be purchased for educational, business, or sales promotional use. For information, please email the Special Markets Department at SPsales@harpercollins.com.

FIRST EDITION

Image credits: p. 1: Marc Porat/Spellbound Productions II; pp. 12, 38, 39, 91, 92, 104, 129, 131, 132, 141, 159, 167, 169, 307: Dwight Eschliman; pp. 28, 30, 97, 139, 152, 244, 245, 246, 272, 275, 278: Matteo Vianello; p. 89: Tony Fadell; p. 101 (top): Manual Creative; p. 101 (bottom): Erik Charlton; p. 142: Will Miller.

Designed by Bonni Leon-Berman
Endpaper illustration by Susan Kare

Library of Congress Cataloging-in-Publication Data has been applied for.

ISBN 978-0-06-304606-1

22 23 24 25 26 LSC 10 9 8 7 6 5 4 3 2 1

To Nana, Gramps, Mom, and Dad—my first mentors

CONTENTS

PART IV: BUILD YOUR BUSINESS

PART V: BUILD YOUR TEAM

PART VI: BE CEO

INTRODUCTION

Many of my experienced, trusted mentors have died.

I looked around a few years ago and the wise, (mostly) patient souls who I had peppered with a million questions, who dealt with my late-night phone calls, who helped me start companies and build products and run board meetings or just be a better person—they were gone. Some much too early.

Now I was the one getting peppered with questions. The same questions I had asked, over and over. Questions about startups, sure, but also more basic stuff: whether to quit a job or not, what career move should I make, how to know if my ideas are any good, how to think about design, how to deal with failure, when and how to start a business.

And weirdly, I had answers. I had advice. I learned it from amazing mentors and the incredible teams I've worked with over thirty years. I learned it at many tiny startups and giant companies, building products that hundreds of millions of people use every day.

So now, if you call me at midnight, panicked, asking how to keep a company's culture intact while it's growing or how not to screw up marketing, I can give you some insights, some tricks and tips, even some rules.

But I won't. Please don't call me at midnight. I've learned the value of a good night's sleep.

Just read this book.

It contains much of the advice I give out daily to new grads and CEOs, to execs and interns, to everyone trying to claw their way through the business world and build their career.

This advice is unorthodox because it's old-school. The religion of Silicon Valley is reinvention, disruption—blowing up old ways of thinking and proposing new ones. But certain things you can't blow up. Human nature doesn't change, regardless of what you're building, where you live, how old you are, how wealthy or not. And over the last thirty-plus years I've seen what humans need to reach their potential, to disrupt what needs disrupting, to forge their own unorthodox path.

So I'm here to write about a leadership style that I've seen win, time and time again. About how my mentors and Steve Jobs did it. About how I do it. About being a troublemaker, a shit-stirrer.

This isn't the only way to make something worth making, but it's my way. And it's not for everyone. I'm not going to be preaching progressive, modern organizational theory. I'm not going to tell you to work two days a week and retire early.

The world is full of mediocre, middle-of-the-road companies creating mediocre, middle-of-the-road crap, but I've spent my entire life chasing after the products and people that strive for excellence. I've been incredibly lucky to learn from the best—from bold, passionate people who made a dent in the world.

I believe everyone should have that chance.

That's why I wrote this book. Everybody trying to do something meaningful needs and deserves to have a mentor and coach—someone who's seen it and done it and can hopefully help you through the toughest moments in your career. A good mentor won't hand you the answers, but they will try to help you see your problem from a new perspective. They'll loan you some of their hard-fought advice so you can discover your own solution.

And it's not just tech entrepreneurs in Silicon Valley who deserve some help. This book is for anyone who wants to create something new, who is chasing excellence, who doesn't want to waste their precious time on this precious planet.

I'm going to talk a lot about building a great product, but a product doesn't have to be a piece of technology. It can be anything you make—a service. A store. It could be a new kind of recycling plant. And even if you're not ready to make anything yet, this advice is still meant for you. Sometimes the first step is just figuring out what you want to do. Getting a job you're excited about. Building the person you want to become or building a team that you can build anything with.

This book isn't trying to be a biography—I'm not dead yet. It's a mentor in a box. It's an advice encyclopedia.

If you're old enough to remember a time before Wikipedia, you might recall the joy of the literal wall of encyclopedias on your bookshelf or your grandparents' study or deep inside the bowels of the library. You'd go to it if you had a specific question, but once in a while you'd also just open it up and start reading. *A* for Aardvark. You'd follow along and see where you ended up, reading straight through or hopping around, discovering little snapshots of the world.

That's how you should think of this book.

- You can read it straight through from beginning to end.
- You can poke around to find the advice and stories that are most interesting or useful in your current career crisis. Because there's always a crisis—either personally, organizationally, or competitively.
- You can follow the "See also" links sprinkled throughout the book just like you'd click through on Wikipedia. Dig deeper into any topic and see where it takes you.

Most business books have one basic thesis that they spend three hundred pages expanding on. If you're looking for a range of good advice on various topics, you might need to read forty books, skimming endlessly to find the occasional nugget of useful information. So for this book I just collected the nuggets. Each chapter has advice and stories informed by the jobs, mentors, coaches, managers, and peers I've had and the countless mistakes I've made.

Since this is my advice based on my experience, this book roughly follows my career. We go from my first job out of college and end up where I am now. Every step, every failure, taught me something. Life didn't begin with the iPod.

But this book isn't about me. Because *I* didn't make anything. I was just one of the people on the teams that made the iPod, iPhone, the Nest Learning Thermostat, and Nest Protect. I was there, but I was never there alone. This book is about what I've learned—typically the hard way.

And to understand the things I've learned, you should probably know a bit about me. So here goes:

1969

The usual start: I was born. And by preschool, we started moving. My dad was a salesman for Levi's and we were always on the road, seeking the next denim gold mine. I went to twelve schools in fifteen years.

1978–1979

Startup #1: Eggs. I sold them door-to-door in third grade. I stand by that company—it was a solid business. I bought eggs cheap from a farmer, then my little brother and I piled them in our blue wagon and pulled it down the neighborhood streets each morning. It gave me pocket money that my parents couldn't tell me how to spend—my first taste of true freedom.

Had I stuck with it, who knows what heights I could have attained.

1980

Found my life's work. It was the summer of fifth grade. A good time to discover your calling. I took a programming class when "programming" meant filling in bubbles with a number 2 pencil on little cards and getting results on a paper printout. There wasn't even a monitor.

It was the most magical thing I had ever seen.

1981

1st love. An Apple][+. Eight bits, a real, gorgeous twelve-inch glowing green monitor, a beautiful brown keyboard.

I had to have that incredible, incredibly expensive machine. My grandfather made me a deal that he'd match whatever money I earned working as a golf caddy, so I worked my ass off until I could afford it.

I cherished that computer. It was my all-abiding passion and my lifeline. By twelve, I'd given up trying to maintain traditional friendships. I knew I'd just move again the next year, so the only way to hold on to friends was through my Apple. There was no internet, no email, but there were 300-baud modems and digital bulletin boards—BBSes in the parlance of the time. I'd find my fellow geeks in whatever school I was in, then we'd keep in touch through our Apples. We taught ourselves to program and hacked the phone companies to get free long-distance calls and bypass the $1–$2/minute fees.

1986

Startup #2: Quality Computers. A friend I connected with over 300-baud started Quality Computers in senior year of high school. I joined him soon after. We were a mail-order company reselling third-party Apple][hardware, DRAM chips, and software from his basement. And we wrote our own software, too—the upgrades and expansion boards we sold were complex to install and harder to use, so we wrote software to simplify everything for mere mortals.

It turned into a real business—an 800 number, warehouses, ads in magazines, employees. A decade later my friend sold it for a couple million. But I was long gone. Selling stuff was okay. Making it was better.

1989

Startup #3: ASIC Enterprises. ASIC stood for Applications Specific Integrated Circuit. I didn't have a ton of branding experience when

I was twenty. But I had a lot of love. In the late eighties my beloved Apple][was struggling. It needed to be faster. So a friend and I decided we were going to save Apple. We built a new, faster processor—the 65816. I did not, in fact, know how to build a processor. I took my first processor design class in college a semester after we started. But we built those chips and they worked eight times faster than what was available—a blazing 33MHz—and even sold some to Apple before they stopped designing new Apple][s.

1990

Startup #4: Constructive Instruments. I teamed up with my professor at the University of Michigan to make a multimedia editor for kids. I threw myself into it, was constantly working or on call. I had a beeper in the days when beepers were exclusively for doctors and drug dealers. The other college kids often asked what was wrong with Fadell—why wasn't he partying and drinking instead of being stuck in a basement, alone with a computer?

By the time I graduated, Constructive Instruments had a few employees. We had an office. A product. Sales partnerships. I was twenty-one years old and CEO. I was also winging it so hard I'm surprised I didn't take off.

1991

Diagnostics Software Engineer at General Magic. I needed to learn how to run a real startup. So I decided to learn from the greats. I got a job at one of the most secretive, exciting companies in Silicon Valley. It was packed to the rafters with geniuses, the opportunity of a lifetime.

We were going to make the most amazing personal communications and entertainment device in history. I drank every last drop of the Kool-Aid and poured my life into that company. We were going to change the world. We couldn't lose.

1994

Lead Software and Hardware Engineer at General Magic. We lost.

1995

CTO at Philips. I started talking to Philips, one of General Magic's partners, about what went wrong. I pitched them my idea: we shift the audience, use existing software and hardware, simplify simplify simplify.

So Philips hired me to make pocket PCs for businesspeople on the go. I became a twenty-five-year-old chief technology officer. It was my second gig out of college.

1997–1998

Launched the Philips Velo and Nino. They were a critical success!

1997–1998

We couldn't sell enough.

1998

Philips Strategy and Ventures Group. I shifted over to the VC side of Philips. Began learning what I could about that world. But the pocket PC bug was still in my brain. Maybe I just hadn't gotten the audience right. Maybe we didn't need to make a PC for businesspeople. Maybe we needed to make a music player for everybody.

1999

RealNetworks. I was going to make a digital music player with the right team, the right tech, the right vision.

1999, six weeks later

I quit. I stepped through the doors and realized the mistake I'd made almost instantly. Real bad juju.

1999

Startup #5: Fuse Systems. Screw it. I'll do it myself.

2000

The dot-com bubble burst. Funding dried up overnight. I did eighty VC pitches. All of them failed. I was desperate to keep my company going.

2001

Apple called. At first I was just hoping to make enough money consulting to keep Fuse open. Then I joined Apple and brought my team with me.

2001, ten months later

We launched the first iPod. It was a critical success!

2001–2006

VP iPod division. After eighteen generations of iPods, we finally got the kinks figured out.

2007–2010

SVP iPod division and iPhone. Then we created the iPhone. My team built the hardware and the foundational software to run and manufacture the phone. Then we launched two more versions. Then I quit.

2010

Took a break. Focused on my family. Got out of the country. Got exactly as far away from work and Silicon Valley as I needed to be.

2010

Startup #6: Nest Labs. Matt Rogers and I started Nest in a garage in Palo Alto. We were going to revolutionize the least sexy product in history: the thermostat. You should have seen people's faces when we told them what our super-secret new startup was going to build.

2011

Launched the Nest Learning Thermostat. It was a critical success! And holy shit. People bought it.

2013

Launched the Nest Protect smoke and carbon monoxide alarm. We were beginning to create an ecosystem—a thoughtful home that could take care of itself and the people inside it.

2014

Google bought Nest for $3.2 billion. Our hardware, Google's software and infrastructure: it was going to be an amazing marriage.

2015–2016

Google created Alphabet. I quit. Nest got kicked out of Google and into Alphabet, which then demanded we drastically change our plans. Then they decided to sell Nest. Not the marriage we signed up for. I walked away in utter frustration.

2010–now

Future Shape. After leaving Google Nest, I focused on some of the advising and investing I'd been doing since 2010. Now we mentor and support around two hundred startups full-time.

My life has swung wildly between success and failure, incredible career highs immediately followed by bitter disappointment. And with each failure I chose to start from scratch, take all that I'd learned and do something completely new, become someone completely new.

The latest version of me is a mentor, coach, investor, and, weirdly now, an author. But being an author only happened because the stars aligned when Dina Lovinsky, a brilliant writer who I worked (and sparred) with for a decade, was available to help and ready to call me on my crap. Young, brash, bold, Dina was there from the earliest days of Nest, watched everything happen firsthand, and learned how to write like I would write, if I could write.

You should probably know now—I'm a terrible writer. I can write software, sure, but write a book? That ain't me. Armed with only a spreadsheet of random lessons learned, I had no idea how to get the first word on the page. But then again, I also had no idea how to build a computer processor, or a music player, or a smartphone, or a thermostat, and those seem to have worked out okay.

The advice in this book is by no means complete, but it's a start. I'm still learning, revising my thoughts each day. Just like everyone else. This book contains some of what I've learned so far.

Part

1

BUILD YOURSELF

I tried to build the iPhone twice.

Everybody knows about the second time. The time we succeeded. Few people know about the first.

In 1989, an Apple employee and intellectual visionary named Marc Porat drew this:

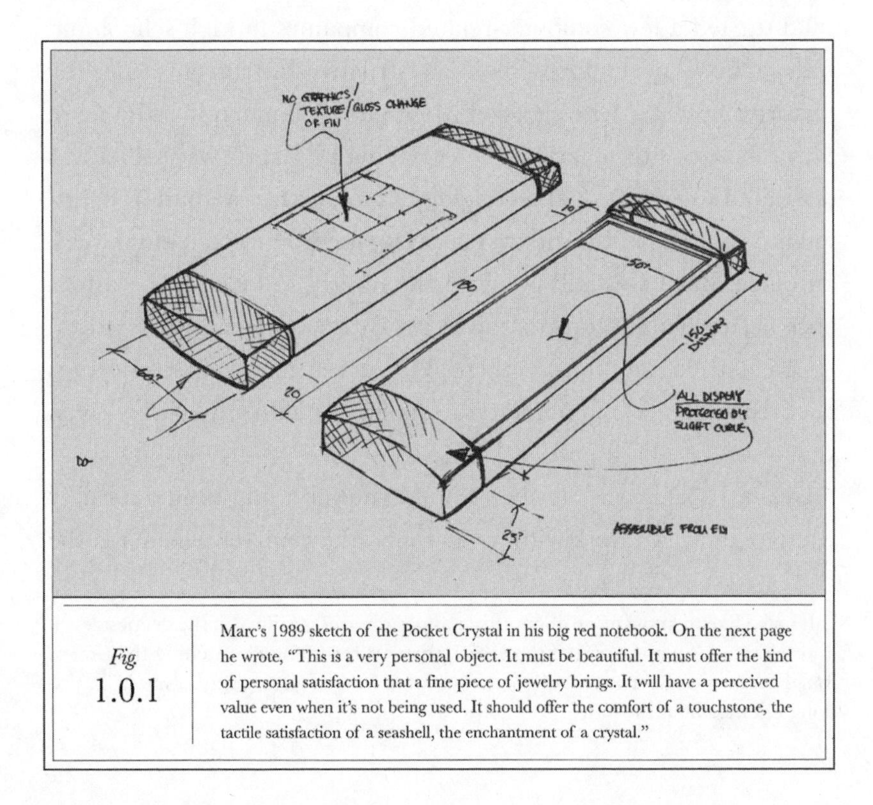

Fig. 1.0.1 Marc's 1989 sketch of the Pocket Crystal in his big red notebook. On the next page he wrote, "This is a very personal object. It must be beautiful. It must offer the kind of personal satisfaction that a fine piece of jewelry brings. It will have a perceived value even when it's not being used. It should offer the comfort of a touchstone, the tactile satisfaction of a seashell, the enchantment of a crystal."

The Pocket Crystal was a beautiful touchscreen mobile computer that combined a cell phone and fax machine, that let you play games and watch movies and buy plane tickets from anywhere.

This insanely prophetic vision was made more completely nuts by the fact that this was—I repeat—1989. The Web didn't exist, mobile gaming meant carrying a Nintendo console to your friend's house, and almost nobody owned—or even really understood the need for—a cell phone. There were pay phones everywhere, everyone's got a pager—why lug around a giant plastic brick with you?

But Marc and two other geniuses and ex-Apple wizards, Bill Atkinson and Andy Hertzfeld, started a company to build the future. They called it General Magic.* I read about it in the "Mac The Knife" rumors section of (the now long-dead) *MacWeek Magazine* right around the time I realized that I had no real idea how to run my startup.

I'd started a few computer-related companies in high school and college, but I'd been focused on Constructive Instruments since my junior year at the University of Michigan. I founded it with one of my professors—the cherubic, "oy vey"-ing Elliot Soloway. Elliot was dedicated to educational technology and together we built a multimedia editor for kids. And we got pretty far—a product, employees, an office. But I was still going to the library to look up the difference between an S-corporation and a C. I was green, green, green. And I had no one to ask—there were no entrepreneurial meetups back then, no Y Combinators. Google wouldn't exist for seven years.

General Magic was my chance to learn everything I could possibly want to know. To work with my heroes—the geniuses who made the

* If you'd like to know more about this company, witness failure at the deepest level, and see that it's not the end of the world, then I'd recommend watching the *General Magic Movie* (www.generalmagicthemovie.com). You might recognize me in it; just don't ask me about the hair.

Apple][, the Lisa, the Macintosh. It was my first real job and my first real chance to change the world like Andy and Bill had.

When I talk to people fresh out of college or early in their careers, that's what they're looking for. An opportunity to make an impact and set themselves on a path to make something great.

But all the stuff they don't and can't teach you in college—how to thrive in the workplace, how to create something amazing, how to deal with managers and eventually become one—it all slaps you in the face the second you step off campus. No matter how much you learn in school, you still need to get the equivalent of a PhD in navigating the rest of the world and building something meaningful. You have to try and fail and learn by doing.

And that means pretty much every young graduate, entrepreneur, and dreamer asks me the same questions:

"What kind of job should I try to get?"

"What kinds of companies should I work for?"

"How do I build a network?"

There's often an assumption that if you find the right job when you're young, you can guarantee some level of success. That your first job out of college connects in a straight line to your second and your third, that at each stage of your career you'll use your inevitable wins to propel yourself upward.

That's what I thought too. I was 100 percent sure General Magic was going to make one of the most impactful devices in history. I poured everything into it. We all did. The team worked literally nonstop for years—we even gave out awards for sleeping in the office for consecutive nights.

Then General Magic imploded. After years of work, tens of millions invested, newspapers shouting that we were destined to beat Microsoft, we sold three to four thousand devices. Maybe five thousand. And that was mostly to family and friends.

The company failed. I failed.

And I spent the next ten years getting kicked in the stomach by Silicon Valley before I made something people actually wanted.

In the process I learned a lot of hard, painful, wonderful, stupid, useful lessons. So for anyone starting their career, or starting a new career, here's what you need to know.

Chapter

1.1

ADULTHOOD

Adulthood is commonly thought of as the time when learning is over and living begins. Yes! I've graduated! I'm done! But learning never ends. School has not prepared you to be successful for the rest of your life. Adulthood is your opportunity to screw up continually until you learn how to screw up a little bit less.

Traditional schooling trains people to think incorrectly about failure. You're taught a subject, you take a test, and if you fail, that's it. You're done. But once you're out of school, there is no book, no test, no grade. And if you fail, you learn. In fact, in most cases, it's the only way to learn—especially if you're creating something the world has never seen before.

So when you're looking at the array of potential careers before you, the correct place to start is this: "What do I want to learn?"

Not "How much money do I want to make?"

Not "What title do I want to have?"

Not "What company has enough name recognition that my mom can brutally crush the other moms when they boast about their kids?"

The best way to find a job you'll love and a career that will eventually make you successful is to follow what you're naturally interested in, then take risks when choosing where to work. Follow your curiosity rather than a business school playbook about how to make money. Assume that for much of your twenties your choices will not work out and the companies you join or start will likely fail.

Early adulthood is about watching your dreams go up in flames and learning as much as you can from the ashes. Do, fail, learn. The rest will follow.

••

I arrived in a cheap, ill-fitting midwestern suit for my interview at General Magic. Everyone was sitting on the floor. They looked up at me, utterly bewildered. Their faces read, "Who is this kid?" They told me to sit down and take off my tie and jacket for Christ's sake.

Mistake #1.

Luckily it was a small one. I became employee #29 in 1991. I was a kid, twenty-one years old, and I gratefully took a job as a diagnostics software engineer. I was going to build software and hardware tools to check other people's designs—the lowest person on the totem pole. But I didn't care. I knew I just needed to get in the door to prove myself and move up.

A month before that, I'd been CEO of my own company. We were tiny—a startup of three, sometimes four people—and were inching along. But it felt more like treading water. And treading water felt like drowning. Either you're growing or you're done. There is no stasis.

So I went where I could grow. The title and the money weren't important. The people were. The mission was. The opportunity was all that mattered.

I remember packing up my stuff to drive to California from Michigan, my belly full of butterflies, four hundred dollars to my name, and my parents trying to understand what the hell was going on.

They wanted me to succeed. They wanted me to be happy. But I truly seemed to be screwing everything up and had been for years. I loved computers, but I'd gotten thrown out of my first computer class in seventh grade nearly every day. I was always telling the teacher he was wrong, always insisting I knew better, never shutting up. I made the poor man cry until they dragged me out of that class and made me learn French instead.

Then I skipped my very first week of college at the University of Michigan to go to Apple Fest in San Francisco and work a booth for my startup. I told my parents after I landed back in Detroit. They were beside themselves. But I'd learned early on to ask for forgiveness, not permission. And I remember the revelation of sitting in my dorm, still digesting the dinner I'd eaten at the wharf in San Francisco, realizing I could be part of two worlds at once. That it wasn't even that hard.

And now I was quitting the company that I had founded, that I had worked day and night to build, that had always seemed like an incredible risk but which was just starting to pay off. And I was going to go—where? General Magic? What the hell was General Magic? If I was going to get a regular job, why not at IBM? Why not Apple? Why not do something stable? Why couldn't I choose a path they could understand?

I wish I'd known this quote then—maybe it would have helped:

> "The only failure in your twenties is inaction. The rest is trial and error."
>
> —Anonymous

I needed to learn. And the best way to do that was to surround myself with people who knew exactly how hard it was to make something great—who had the scars to prove it. And if it turned out to be the wrong move, well, making a mistake is the best way to not make that mistake again. Do, fail, learn.

The critical thing is to have a goal. To strive for something big and hard and important to you. Then every step you take toward that goal, even if it's a stumble, moves you forward.

And you can't skip a step—you can't just have the answers handed to you and detour around the hard stuff. Humans learn through productive struggle, by trying it themselves and screwing up and doing it differently next time. In early adulthood you have to learn to embrace that—to know that the risks might not pan out but to take them anyway. You can get guidance and advice, you can choose a path by following someone else's example, but you won't really learn until you start walking down that path yourself and seeing where it takes you.

I give a speech at high schools sometimes—at graduations where a bunch of eighteen-year-old kids are heading out into the world, alone, for the first time.

I tell them that they probably make 25 percent of their decisions. If that.

From the moment you're born until you move out of your parents' house, almost all your choices are made, shaped, or influenced by your parents.

And I'm not just talking about the obvious decisions—which classes to take, which sports to play. I mean the millions of hidden decisions you'll discover when you leave home and start doing things for yourself:

What type of toothpaste do you use?

What kind of toilet paper?

Where do you put the silverware?

How do you arrange your clothes?

What religion do you follow?

All these subtle things that you never made a decision about growing up are already implanted in your brain.

Most kids don't consciously examine any of these choices. They mimic their parents. And when you're a kid, that's usually fine. It's necessary.

But you're not a kid anymore.

And after you move out of your parents' house, there's a window—a brief, shining, incredible window—where your decisions are yours alone. You're not beholden to anyone—not a spouse, not kids, not parents. You're free. Free to choose whatever you'd like.

That is the time to be bold.

Where are you going to live?

Where are you going to work?

Who are you going to be?

Your parents will always have suggestions for you—feel free to take them or ignore them. Their judgment is colored by what they want for you (the best, of course, only the best). You'll need to find other people—other mentors—to give you useful advice. A teacher or cousin or an aunt or the older kid of a close family friend. Just because you're on your own doesn't mean you have to be alone with your decisions.

Because this is it. This is your window. This is your time to take risks.

When you're in your thirties and forties, the window begins to close for most people. Your decisions can no longer be entirely your own. That's okay, too—great even—but it's different. The people who depend on you will shape and influence your choices. Even if you don't have a family to support, you'll still accumulate just a little more each year—friends, assets, social standing—that you won't want to risk.

But when you're early in your career—and early in your life—the worst that can happen if you take big risks is probably moving back with your parents. And that is not shameful. Throwing yourself out there and having everything blow up in your face is the world's best way to learn fast and figure out what you want to do next.

You might screw up. Your company might fail. You might have so many butterflies in your stomach you'll be worried you got food poisoning. And that's okay. It's exactly what should be happening. If you don't feel those butterflies then you're not doing it right. You have to push yourself up the mountain, even if it means you might fall off a cliff.

I learned more from my first colossal failure than I ever did from my first success.

General Magic was an experiment. Not just in what we were making—and we were making something wholly, ridiculously, almost unbelievably new—but also in how to structure a company. The team was so impressive, packed with genius after genius, that there was no regard for "management." No defined process. We just kind of . . . made stuff. Whatever our leaders thought would be cool.

And every piece had to be hand-built, from scratch. It was like giving one hundred artisans a pile of sheet metal, plastic, and glass and telling them to build a car. One of my projects was to figure out how to connect various gadgets into our device, so I built the precursor to the USB port. Then I was assigned to build an infrared network to work between devices (like how a remote connects to a TV)—so I reinvented all seven layers of a protocol stack. Amazingly, I made it work. The other engineers were excited and created a word game on top of it. The game became a hit around the office. I was ecstatic, over the moon. But eventually a more experienced engineer got around to looking at what I'd coded and, baffled, asked why I'd built a network protocol that way. I answered that I didn't know I was building a network protocol.

Mistake #2.

But even though I could have just read a book and saved myself days of work—man, did it feel good. I'd made something the world had never seen before, something useful, and I made it my way.

It was crazy. But it was fun. Especially in the beginning when everybody was focused on fun. There was no dress code. No rules for the office. It was so different from what I was used to in the Midwest. General Magic was probably one of the first Silicon Valley companies that truly embodied the idea that playing at work was worthwhile—that a joyful workplace could make a joyful product.

And we probably took the joy a little far. Once we were in the office in the middle of the night, working late as usual, and I grabbed the three-person slingshot (doesn't everyone have a slingshot in their office?). Two accomplices and I loaded it with slime, fired, and punched a giant hole through a large third-story window. I was terrified I'd get fired.

Everyone just laughed.

That was Mistake #3.

For four years, I threw myself into General Magic. I learned and screwed up and worked and worked and worked. Ninety, 100, 120 hours a week. I never liked coffee, so I survived primarily off Diet Coke. A dozen a day (for the record, I haven't touched the poison since).

(I don't recommend working that much, by the way. You should never kill yourself for your job, and no job should ever expect that of you. But if you want to prove yourself, to learn as much as you can and do as much as you can, you need to put in the time. Stay late. Come in early. Work over the weekend and holidays sometimes. Don't expect a vacation every couple of months. Let the scales tip a little on your work/life balance—let your passion for what you're building drive you.)

For years I ran full tilt in whatever direction people pointed me in—and we were going in every direction at once. My heroes would say take that hill, and by God, I would make it my Everest and do whatever it took to impress them. I was 100 percent sure we were going to make the most world-changing device in history. We all were.

Then the launch got delayed. Again. And again. And again. We had no shortage of money or press or all the sky-high expectations to go with it, so the product just kept growing. Never quite good enough or done enough. Our competitors started popping out of the woodwork. We were creating a private network system run by major telecom companies like AT&T just as the internet began to go mainstream, open to everyone. Our processor didn't have enough oomph to support the ambitious user experience (UX) that Andy and Bill dreamed up nor the graphics and icons that Susan Kare designed. Susan is a brilliant artist who created the original visual language for the Mac, and she made an entire beautiful world for the Magic Link. But then every time you tapped on the screen, the damn thing would freeze. User testers were frustrated by the waiting, the bugs, never knowing if they'd done something wrong or the device had just stopped working. The list of issues would get longer each day.

Mistake #4 through mistake #4000.

When we finally shipped in 1994, we hadn't made the Pocket Crystal. We made the Sony Magic Link.

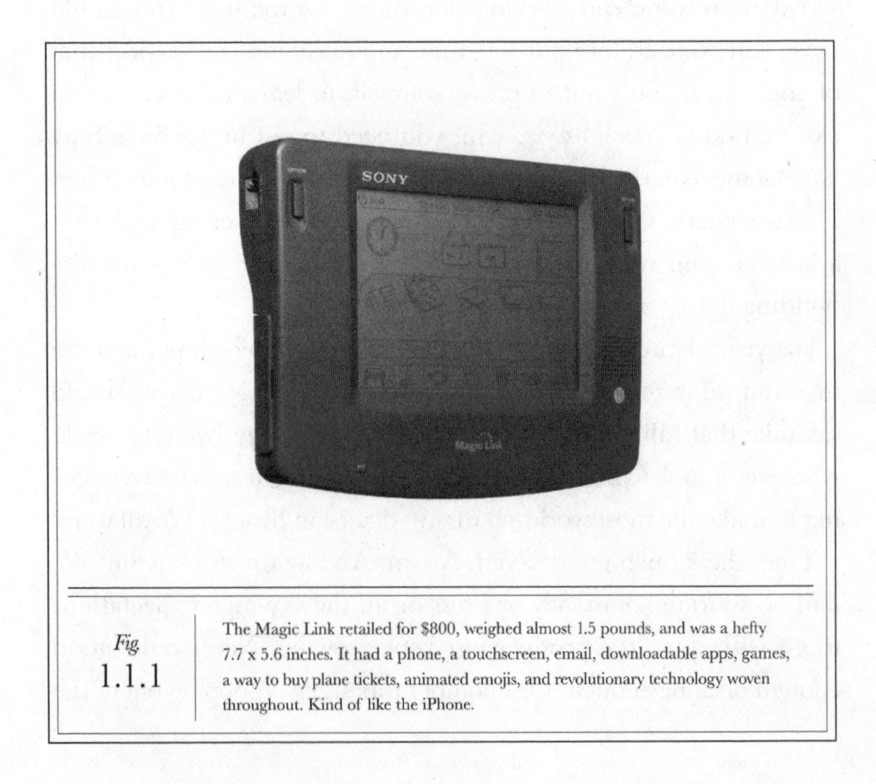

Fig 1.1.1 The Magic Link retailed for $800, weighed almost 1.5 pounds, and was a hefty 7.7 x 5.6 inches. It had a phone, a touchscreen, email, downloadable apps, games, a way to buy plane tickets, animated emojis, and revolutionary technology woven throughout. Kind of like the iPhone.

It was deeply flawed and strangely suspended between the past and the future—it had both animated emojis and a little printer for faxes. But it was still absolutely, flat-out, ahead-of-its-time amazing. A first step into a different world, where everyone could carry a computer anywhere. All the work, the lack of sleep, the toll on my body, the toll on my parents—it was going to be worth it. I was incredibly proud of it. I was so thrilled with what our team had created. I still am.

And then nobody bought it.

After spending all those days and nights in the office, I woke up and couldn't get out of bed. I felt it in my chest. Everything we had done ended in failure. Everything.

And I ultimately knew why.

By the time General Magic was unraveling around me, I wasn't just a lowly diagnostics engineer anymore. I'd worked on silicon, hardware, and software architecture and design. When things started to go awry, I'd ventured out and started talking to people in sales and marketing, began learning about psychographics and branding, finally grasped the importance of managers, of process, of limits. After four years, I realized there was a whole world of thinking that was needed before a line of code should be written. And that thinking was fascinating. That thinking was what I wanted to do.

The enormous gut punch of our failure, of my failure, of everything I'd worked for falling apart—it made the path in front of me strangely clear: General Magic was making incredible technology but wasn't making a product that would solve real people's problems. But I thought I could.

That's what you're looking for when you're young, when you think you know everything then suddenly realize you have no idea what you're doing: a place where you can work as hard as you can to learn as much as you can from people who can make something great. So even if the experience kicks your ass, the force of that kick will propel you into a new stage of your life. And you'll figure out what to do next.

Chapter

1.2

GET A JOB

If you're going to throw your time, energy, and youth at a company, try to join one that's not just making a better mousetrap. Find a business that's starting a revolution. A company that's likely to make a substantial change in the status quo has the following characteristics:

1. It's creating a product or service that's wholly new or combines existing technology in a novel way that the competition can't make or even understand.

2. This product solves a problem—a real pain point—that a lot of customers experience daily. There should be an existing large market.

3. The novel technology can deliver on the company vision—not just within the product but also the infrastructure, platforms, and systems that support it.

4. Leadership is not dogmatic about what the solution looks like and is willing to adapt to their customers' needs.

5. It's thinking about a problem or a customer need in a way you've never heard before, but which makes perfect sense once you hear it.

••

Cool technology isn't enough. A great team isn't enough. Plenty of funding isn't enough. Too many people throw themselves blindly at hot trends, anticipating a gold rush, and end up falling off a cliff. Look at the body count of virtual reality (VR)—dead startups as far as the eye can see and billions of dollars burned up over the past thirty years.

"If you make it, they will come" doesn't always work. If the technology isn't ready, they won't come for sure. But even if you've got the tech, then you still have to time it right. The world has to be ready to want it. Customers need to see that your product solves a real problem they have today—not one that they may have in some distant future.

I think of this as the General Magic problem. We were trying to build an iPhone years before it was a glimmer in Steve Jobs's eye.

And you know what utterly beat us? Palm. Because Palm PDAs let you put the phone numbers you kept on scraps of paper or on your desktop computer into a device that you could carry with you. That's it. That simple. You couldn't jam a Rolodex into your pocket or purse, so Palm was the right solution for the time. It made sense. It had a reason to exist.

General Magic did not. We started from the technology—focusing on what we could create, what would impress the geniuses at our company—not the reason why real, nontechnical people would need it. So the Magic Link solved problems that regular people wouldn't recognize for more than a decade. And because nobody else was building technology for nonexistent problems, the networks, processors, and input mechanisms our products depended on weren't good enough. We had to make everything ourselves. Magic CAP, a revolutionary object-oriented operating system. TeleScript, a new client server programming language. We created servers with online applications and stores. And ultimately, even though it fell short of the vision, we built something really incredible. For us geeks.

For everyone else, it was kind of neat. Maybe. If they even understood what it was. A luxury toy for rich people or nerds or very rich nerds. A plaything.

If you're not solving a real problem, you can't start a revolution.

A glaring example is Google Glass or Magic Leap—all the money and PR in the world can't change the fact that augmented reality (AR) glasses are a technology in search of a problem to solve. There's just no reason for the general public to buy them. Not yet. Nobody can quite imagine walking into a party or the office with these weird ugly glasses on their face, creepily filming everyone around them. And even if there's a brilliant vision for the future of AR glasses, the technology can't deliver it yet and the social stigma will take a long time to dissolve. I'm convinced it'll happen, but they're still years away.

On the other hand, take Uber. The founders started with a customer problem—a problem they experienced in their daily lives—then applied technology. The problem was simple: finding a cab in Paris was next to impossible and hiring private drivers was expensive and took forever. In the days before smartphones, the solution might have been to simply start a new kind of taxi or limo business. But the company's timing was perfect—the sudden ubiquity of smartphones provided Uber with a platform and put customers into the right mindset to accept their solution. If I can order a toaster with an app on my phone, why shouldn't I hail a car? That combination of a real problem, the right timing, and innovative technology allowed Uber to shift the paradigm—to create something that traditional cab companies couldn't even dream of, never mind compete with.

And this isn't just a Silicon Valley phenomenon. Revolutionary companies are popping up in every industry—in agriculture, drug discovery and creation, finance and insurance—in every part of the world. Seemingly impossible problems that a decade ago would have cost billions to solve, requiring massive investments from giant firms, can now be figured out with a smartphone app, a small sensor, and the internet. And that means there are thousands of people all over the world finding opportunities to change the way people work and live and think.

Take whatever job you can at one of those companies. Don't worry too much about the title—focus on the work. If you get a foot in the door at a growing company, you'll find opportunities to grow, too.

Just whatever you do, don't become a "management consultant" at a behemoth like McKinsey or Bain or one of the other eight consultancies that dominate the industry. They all have thousands upon thousands of employees and work almost exclusively with Fortune 5000 companies. These corporations, typically led by tentative, risk-averse CEOs, call in the management consultants to do a massive audit, find the flaws, and present leadership with a new plan that will magically "fix" everything. What a fairy tale—don't get me started.

But to many new grads, it sounds perfect: you get paid incredibly well to travel around the world, work with powerful companies and executives, and learn exactly how to make a business successful. It's an alluring promise.

Parts of it are even true. Yes, you get a nice paycheck. And yes, you get plenty of practice pitching important clients. But you don't learn how to build or run a company. Not really.

Steve Jobs once said of management consulting, "You do get a broad cut at companies but it's very thin. It's like a picture of a banana: you might get a very accurate picture but it's only two dimensions, and without the experience of actually doing it you never get three dimensional. So you might have a lot of pictures on your walls, you can show it off to your friends—I've worked in bananas, I've worked in peaches, I've worked in grapes—but you never really taste it."

If you do choose to go that route and find yourself at one of the Big Four or the other top six firms, then that is of course your choice. Just know before you go what you want to learn and the experiences you need for your next chapter. Don't get stuck. Management consulting should never be your endpoint—it should be a way station, a brief pause on your journey to actually doing something. Making something.

To do great things, to really learn, you can't shout suggestions from the rooftop then move on while someone else does the work. You have to get your hands dirty. You have to care about every step, lovingly craft every detail. You have to be there when it falls apart so you can put it back together.

You have to actually do the job. You have to love the job.

But what happens if you fall in love with the wrong thing? If you find a product or company that's too early—the supporting infrastructure isn't there, the customers don't exist, the leadership has a crazy vision and won't budge.

What if you're deeply passionate about quantum computing or synthetic biology or fusion energy or space exploration even though there's no sign that any of those industries will bear fruit anytime soon?

Then screw it. Go for it. If you love it, don't worry about all my advice, don't worry about the timing.

I spent the dot-com bubble building handheld devices. After General Magic started floundering, the obvious solution was to jump ship to Yahoo or eBay and join the internet gold rush. That's what everyone told me to do. "Are you crazy, why go to Philips?! The internet is where all the money is! No one needs more consumer computing devices."

But I went to Philips anyway. I knew there was room for something amazing between desktop computers and cell phones. I saw it, felt it, when I was at General Magic. So I built a team to make devices at Philips and then started my own company to make digital music players. I stuck with it because I loved it—loved building the whole system from the bottom up, the atoms and electrons, the hardware and software and networks and design. And by the time Apple called me to make the iPod, I knew exactly how to do it.

If you're passionate about something—something that could be solving a huge problem one day—then stick with it.

Look around and find the community of people who are passionate about it, too. If there's nobody else on Earth thinking about it, then you may truly be too early or going in the wrong direction. But if you can find even a handful of like-minded people, even if it's just a tiny community of geeks building technology nobody has any idea how to turn into a real business, then keep going. Get in on the ground floor, make friends, and find mentors and connections that will bear fruit when the world spins just enough to make what you're making make sense. You may not be at the same company as when you started, the vision may be different, the product may be different, and the technology will have changed. You may have to fail and fail and learn and learn and evolve and understand and grow.

But one day, if you are truly solving a real problem, when the world is ready to want it, you'll already be there.

What you do matters. Where you work matters. Most importantly, who you work with and learn from matters. Too many people see work as a means to an end, as a way to make enough money to stop working. But getting a job is your opportunity to make a dent in the world. To put your focus and energy and your precious, precious time toward something meaningful. You don't have to be an executive right away, you don't have to get a job at the most amazing, world-changing company right out of college, but you should have a goal. You should know where you want to go, who you want to work with, what you want to learn, who you want to become. And from there, hopefully you'll start to understand how to build what you want to build.

Chapter

1.3

HEROES

Students seek out the best professors on the best projects when getting their master's or PhD, but when they look for jobs, they focus on money, perks, and titles. However, the only thing that can make a job truly amazing or a complete waste of time is the people. Focus on understanding your field and use that knowledge to create connections with the best of the best, people you truly respect. Your heroes. Those (typically humble) rock stars will lead you to the career you want.

••

If there are gods of software design and coding, they are Bill Atkinson and Andy Hertzfeld. Their faces were in the magazines I'd read religiously cover to cover since grade school. I'd used everything they'd ever built—the revolutionary Mac, MacPaint, Hypercard, the Lisa.

They were my heroes. When I met them, I felt like I was meeting the president. The Beatles. Led Zeppelin. My palms were sweating when I shook their hands; I could barely catch my breath. But after time passed and I blinked the stars out of my eyes, I realized they were approachable, easy to talk to—a rare trait in the world of geniuses. And I could talk to them for hours. About coding, about design and UX, about a million things I was curious about. I even showed them the product my startup Constructive Instruments had created.

That's the main reason I got the job at General Magic, I think, even though there were people sleeping on their doorstep to get an interview and I was a nobody geek from Michigan. Not because I could flatter the founders or because I hounded the HR director before and after the interview (although in that pre-email era I did call her every day for a month—before and after the interview), but because I had gained a huge amount of practical, useful information through sheer brute force. I spent most of my time building—chips and software and devices and companies—and the rest of my time reading everything I could get my hands on about the industry. And that's what set me apart. That's what can set anyone apart. Bill Gurley, the incredibly smart, wry, contrarian Silicon Valley VC and Texan deal maker, puts it this way: "I can't make you the smartest or the brightest, but it's doable to be the most knowledgeable. It's possible to gather more information than somebody else."

And if you're going to devote that much time to gathering information, then learn about something you'd be interested in even if you weren't trying to get a job doing it. Follow your curiosity. Once you're armed with that knowledge, then you can start hunting down the people who are the best of the best and trying to work with them. And that doesn't mean stalking Elon Musk if you're into electric cars. Look at who reports to him. And who reports to them. And which competing company would kill to hire those people. Understand the subdisciplines and see who leads the one you're most interested in. Find the experts on Twitter or YouTube, then send them a message, a comment, a LinkedIn connection. You're interested in the same things, you have the same passions—so share your point of view, ask a smart question, or just tell them about some fascinating minutiae that your family and friends find deeply, desperately, unfathomably boring.

Make a connection. That's the best way to get a job anywhere.

And if that seems impossible—if you follow your heroes on Twitter but can't imagine they'll ever pay any attention to you—I'm excited to tell you that that is complete bullshit. I doubt I'm anyone's hero, but I'm an experienced, well-connected product designer who's been lucky enough to make some famous technology. Most people assume I won't pay any attention to people who randomly DM me on Twitter or send me unsolicited emails out of the blue. But sometimes I do.

Not when people are just asking for a job. Or angling for funding. But I'll notice people who come with something interesting to share. Something smart. Especially if they keep coming. If they sent me something cool last week and something cool this week and they keep bringing fascinating news or tech or ideas and they're persistent, then I'll start to recognize them. I'll start to remember them, and respond. And that can turn into an introduction or a friendship or a referral or, potentially, a job at one of our portfolio companies.

The key is persistence and being helpful. Not just asking for something, but offering something. You always have something to offer if you're curious and engaged. You can always trade and barter good ideas; you can always be kind and find a way to help.

Look at Harry Stebbings. He's a smart, genuine, incredibly nice person who started the *20 Minute VC* podcast in 2015. Then he just started inviting people to be on it. He put himself out there. He was persistent. He was helpful and warm. And he started getting some traction—first one CEO, then another, founders and investors and high-level execs. Including me—it was one of my favorite podcast interviews.

After every podcast, he asks his interviewee in private, "Who are the top three people you know and respect who you think I should talk to next? Do you mind intro-ing me really quickly?"

In 2020, he was able to parlay his success and network into a small VC fund. In 2021, that fund got an additional $140 million in funding.

As I write this, Harry Stebbings is twenty-four years old.

And I'm not saying that every tweet or LinkedIn message to your heroes is going to turn into a $140 million VC fund. But it might turn into a job. It might even turn into a job with your heroes.

And any job working with your heroes is a good job.

But if you can, try to get into a small company. The sweet spot is a business of 30–100 people building something worth building, with a few rock stars you can learn from even if you aren't working with them every day.

You could go to Google, Apple, Facebook, or some other giant company, but it'll be hard to maneuver yourself to work closely with the rock stars. And you should know you're not going to make a real impact. Not for a long time. You're a pebble bouncing off an elephant. But you'll be a well-paid pebble eating free kale chips, so if you do go that route enjoy the paycheck while working on your tiny piece of some vast and endless project. Then spend your ample free time getting a feel for the structures and divisions, the micro-disciplines, the processes, the research, the long-term projects and long-term thinking a company can do when shipping tomorrow isn't critical to its survival. That stuff is good to know. [See also: Chapter 4.2: Are You Ready?] But don't get stuck between the elephant's toes so you can never see the whole beast. It's easy to mistake navigating processes, red tape, job leveling, and politics for real personal growth.

A small company has fewer resources, less equipment, tiny budgets. It may not be a success and may never make any money. It may not have many perks (although that could be a good thing). Startups that blow all their funding on Ping-Pong instructors and free beer aren't focused on the right things. [See also: Chapter 6.4: Fuck Massages.] But you'll work with a wider range of talented people—in sales, marketing, product, operations, legal, maybe even quality or customer support. Smaller companies still have specialization, but usually without silos. And they have a different energy. The whole company will be focused on working together to make one precious idea become reality. Anything unnecessary is shunned; red tape and politics are typically nonexistent. There's more riding on what you do because it's actually meaningful to the company surviving. You're all in the lifeboat together.

And being in that lifeboat with people you deeply respect is a joy. It is the best time you can have at work. It might be the best time you can have, period. And it doesn't have to end once you're on dry land.

Among the many amazing people who I got to work with at General Magic were Wendell and Brian Sander. Father and son, both insanely brilliant, salt of the earth, engineer's engineers. Brian was my manager at General Magic and both the Sanders helped me figure out how to create MagicBus, a digital peripheral bus for the Magic Link. The ideas and patents we created together are now the basis for USB devices around the world. It was a dream come true.

When General Magic imploded, we all went our separate ways. But I never lost touch. And ten years later, I hired Brian to work on the iPod with me. And then Brian hired his dad.

Once Wendell and I were walking into Apple's main building, Infinite Loop 1, and we bumped into Steve Jobs. Steve was thrilled. Wendell had been Apple employee #16 but Steve hadn't seen him in years. "Wendell! Where you at these days?"

And Wendell said, "I'm here. Working on the iPod. With Fadell."

When you get a chance to work with legends and heroes and gods, you realize they're none of the things that you've fabricated in your brain. They can be geniuses in one area and clueless in another. They can raise you up by praising your work, but you can also help them, catch things they miss, and build a relationship based not on starry-eyed hero worship but mutual respect.

And let me tell you, there is nothing in the world that feels better than helping your hero in a meaningful way and earning their trust—watching them realize you know what you're talking about, that you can be relied on, that you're someone to remember. And then seeing how that respect evolves as you move on to the next job, and the next.

That's the great thing about heroes. You can use their inspiration to drive you. If you do it right, and listen carefully, they'll share decades of learning. And then, one day, you might return the favor.

Chapter

1.4

DON'T (ONLY) LOOK DOWN

The job of an individual contributor (IC)—a person who doesn't manage others—is usually to craft something that needs to be completed that day or in the next week or two. Their responsibility is to sweat the details, so most individual contributors depend on their managers and executive team to set a destination and lay out a path for them so they can keep their focus on the work.

However, if an IC is constantly looking down, their eyes exclusively on their own tight deadlines and the minutiae of their job, they may walk directly into a brick wall.

As an IC, you need to occasionally do two things:

1. **Look up:** Look beyond the next deadline or project and forward to all the milestones coming up in the next few months. Then look all the way down to your ultimate goal: the mission. Ideally it should be the reason you joined the project in the first place. As your project progresses, be sure the mission still makes sense to you and that the path to reach it seems achievable.

2. **Look around:** Get out of your comfort zone and away from the immediate team you're on. Talk to the other functions in your company to understand their perspectives, needs, and concerns. This internal networking is always useful and it can give you an early warning if your project is not headed in the right direction.

••

I only looked up when the sky was falling. Really falling. Before that, an occasional asteroid would rocket through my cubicle at General Magic—when the parts for a good touchscreen hadn't been invented yet, or the software I just wrote "broke the build," or the mobile networks we needed barely functioned—but I'd just brush off my keyboard and keep going.

I trusted Bill, Andy, and Marc to steer the ship. All that was required of me was to prove myself. That's the one downside of working with your heroes. You're so busy learning your craft from them that you just assume they're looking at the big picture. You assume they'll notice the brick wall directly in their path.

Think of a project as a straight line in time—there's a beginning and (hopefully) an end. Everyone is walking at the same pace, day by day, on parallel lines—a line for engineering, marketing, sales, PR, customer support, manufacturing, legal, etc.

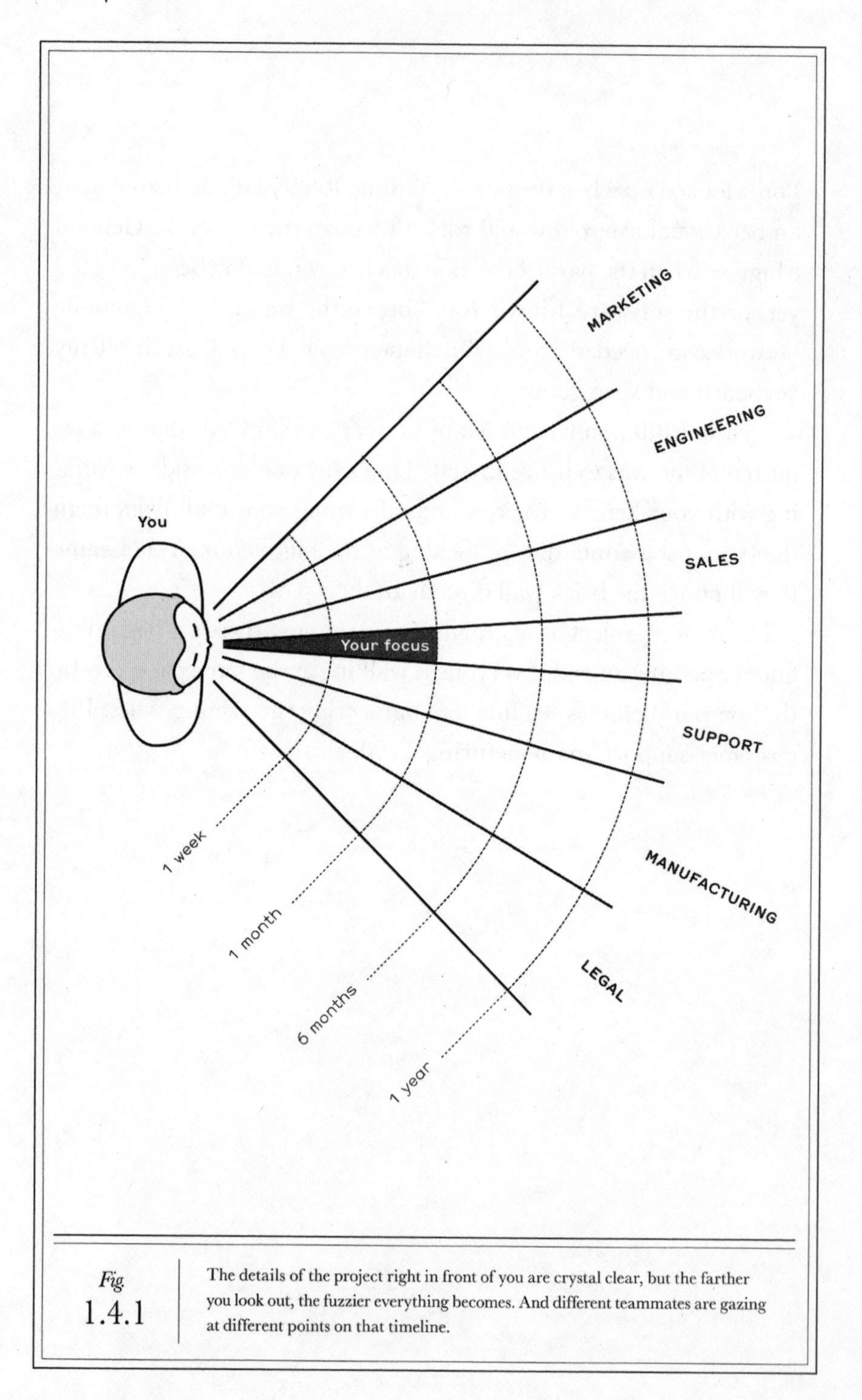

Fig. 1.4.1 The details of the project right in front of you are crystal clear, but the farther you look out, the fuzzier everything becomes. And different teammates are gazing at different points on that timeline.

The CEO and executive team are mostly staring way out on the horizon—50 percent of their time is spent planning for a fuzzy, distant future months or years away, 25 percent is focused on upcoming milestones in the next month or two, and the last 25 percent is spent putting out fires happening right now at their feet. They also look at all the parallel lines to make sure everyone is keeping up and going in the same direction.

Managers usually keep their eyes focused 2–6 weeks out. Those projects are pretty fleshed out and detailed, though they still have some fuzzy bits around the edges. Managers' heads should be on a swivel—they often look down, sometimes look further out, and spend a fair amount of time looking side to side, checking in on other teams, making sure everything's coming together for the next milestone.

Junior individual contributors spend 80 percent of their time looking straight down—maybe a week or two out—to see the fine points of their day-to-day work. In the early stages of your career, that's the way it should be. You should be focused on getting your specific piece of each project done, done well, and out the door.

Your executive team and managers are supposed to be looking out for roadblocks. They're supposed to warn you so you can adjust course, or at least grab a helmet.

But sometimes they don't.

So 20 percent of the time, individual contributors need to look up. And they need to look around. The sooner they start, the faster and higher they'll advance in their career.

Your job isn't just doing your job. It's also to think like your manager or CEO. You need to understand the ultimate goal, even if it's so far away that you're not really sure what it'll look like when you get there. That's helpful in your day-to-day—knowing your destination lets you self-prioritize and make decisions about what you're doing and how you're doing it. But it's also bigger than that. You want to make sure the direction you're headed in still feels right—that you still believe in it.

And you can't ignore the other teams who are working by your side.

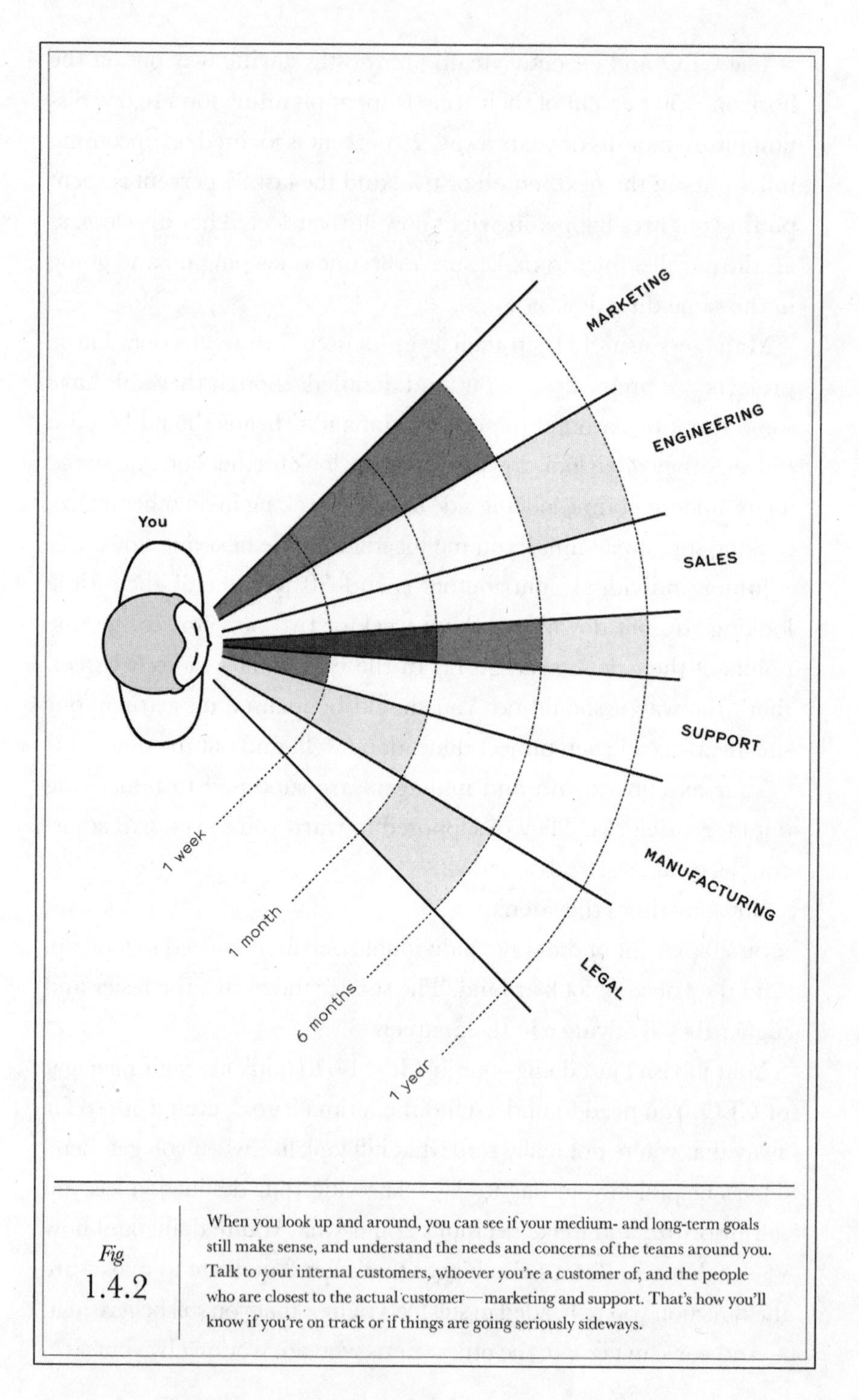

Fig. 1.4.2

When you look up and around, you can see if your medium- and long-term goals still make sense, and understand the needs and concerns of the teams around you. Talk to your internal customers, whoever you're a customer of, and the people who are closest to the actual customer—marketing and support. That's how you'll know if you're on track or if things are going seriously sideways.

The first time an asteroid hit me squarely in the face was when I was having lunch with Tracy Beiers. Tracy's a true prankster, an original. As a product manager and an experienced marketer from Microsoft, she'd seen it all when she was working on Windows 1.0.

"I don't know why anyone needs the lemon," she said. She was talking about a little animated emoji we'd just added. It would walk across your emails, doing things even modern emojis can't dream of. Ah, I thought. She wasn't an engineer. She didn't get it. So I rushed to explain: it was so innovative! Here's how we made it! So cool, right? Didn't she think so?

"Yeah, it's cute, I guess," she said, and shrugged. "But I just want email that works. I don't care about the lemon. Nobody's going to care about the damn walking lemon."

Huh. Everyone on the engineering team loved it. So I said, "Tell me more."

I had never approached the product from her point of view. She forced me to take off my engineering, techno-savvy rose-colored glasses and look at what we were building from the perspective of a regular human being.

It was a hard conversation—I was puzzled, dumbfounded. But it was also incredibly useful for both of us. I wanted to understand her perspective. She wanted to understand my side of the business. Mostly, she wanted to know what the hell I was doing.

Not only was she worried that the features we were building were charming but useless, she was also worried we wouldn't actually build them.

"We just worked with Sony marketing to make an ad campaign saying the Magic Link will be able to do all these things. Is it true? Can we actually deliver?"

This was probably around the fifth time we'd pushed back our launch date. Many of the features we'd promised investors and partners had fallen through. The product was slow and buggy. And she wanted to know what was happening behind the scenes—not just what she was hearing from leadership.

Where will wireless messaging work? Where won't it work? What's the customer experience really going to be? What are the trade-offs?

I told her. Then I asked what she thought. And that's when I noticed the sky was falling.

I hadn't realized it, but all those people working parallel to me could see things I couldn't. They had a completely different view of our world—a view that I wanted to understand.

New perspectives are everywhere. You don't have to drag a bunch of people off the street to stare at your product and tell you what they think. Start with your internal customers. Everyone in a company has customers, even if they're not building anything. You're always making something for someone—the creative team is making stuff for marketing, marketing is making stuff for the app designers, the app designers are making stuff for the engineers—every single person in the company is doing something for someone, even if it's just a coworker on another team.

You're somebody's customer, too—so talk to whoever is doing work for you. Show up with something of value or a pertinent question. Try to understand what their roadblocks are and what they're excited about.

And talk to the people who are closest to the customer, like marketing and support—find teams who communicate with customers day in and day out and hear their feedback directly.

Come curious. And come genuinely interested. When you're looking up and around, you're not on a self-serving mission to understand if your company will fail and how quickly you should cut and run. You're trying to understand how to do your job better. You're getting ideas of how to help your project and your company's mission succeed. You're starting to think like your manager or leader, which is the first step to becoming a manager or leader.

And as you do that, you may start seeing things emerge from the haze.

Hard things. Brick-wall-like things.

When I finally looked up and around, I realized we were bashing our heads against a wall that was never going to move. Our mission was still inspiring, but our path there was blocked. So I held on to the mission, but I found a new goal to walk toward. I stepped off the path and took a hard left. And that's how I found my next job.

The most wonderful part of building something together with a team is that you're walking side by side with other people. You're all looking at your feet and scanning the horizon at the same time. Some people will see things you can't, and you'll see things that are invisible to everyone else. So don't think doing the work just means locking yourself in a room—a huge part of it is walking with your team. The work is reaching your destination together. Or finding a new destination and bringing your team with you.

Part

II

BUILD YOUR CAREER

I wanted to save General Magic.

When it became cruelly clear that no one but our close, geeky friends was going to buy the Magic Link, I nervously pitched an idea to my heroes: let's pivot. Instead of making a communication and entertainment device for the general public, let's focus on businesspeople.

General Magic's target customer was "Joe Sixpack." Seriously. It's a derogatory caricature of an average American slumped on his couch, drinking beer, watching football, not thinking about much of anything. It's a terrible way to imagine your customer. And even though we repeated it over and over, claimed we were doing it all for him—it was meaningless. Because even if Joe Sixpack existed, he was never going to buy the Magic Link. This was before the internet was remotely accessible, at a time when most people didn't have a desktop computer, didn't have email, couldn't imagine mobile games or movies.

It was 1992. Joe Sixpack had literally no reason to put a computer in his pocket.

But businesspeople just might. They were beginning to use email and notes and digital calendars. They needed all their contacts on a mobile device instead of on a ten-pound laptop. They were like my dad—always on the move from city to city. And they were constantly rushing out of cars and planes to feed quarters into pay phones to listen to their voicemails, trying to make deals and catch meetings in an era before cell phones. They had a problem we could solve.

It was so glaringly, wonderfully obvious. Make a product for people who already saw the need and felt the pain daily. Just take the Magic Link and put a keyboard on it. Strip out the quirks and whimsy—kill the walking lemons—and focus on work. Create a mobile business-oriented device, user interface, and suite of applications. Add word processing and spreadsheets. I started talking to people at General Magic, trying to get them interested. First I pitched it to my peers, then the leadership team.

"Good idea, but . . ." they said. We went back and forth for a while, everyone trying to be nice. Trying to make it happen. But in the end the answer was a resounding "No." It was too much work—we'd have to change too much. We can't do it now. We have other priorities.

But Philips was in. As a major partner and investor in General Magic, they were already building some semiconductors and processor parts for us, so it was easy to reach out to them. And they liked the idea: build a business-focused pocket computer within Philips, but use General Magic hardware and software. I could keep the dream alive and Philips could keep itself relevant in the emerging world of software-driven devices.

So in 1995 I closed the door on the General Magic office where we raced remote control cars through the hallways and pranked each other with hidden hot dogs in the ceiling and walked into a whole different world. I knew Philips would be different, but it was shocking.

Dark wood paneling with smoke stains from the 1970s, no cubicles, constant meetings, managers who just said no to everything. An old guard of old Dutch guys complaining about the lack of Douwe Egberts coffee and frikandel (if you don't know, don't ask). Everywhere I looked I saw the same bad suit I'd ditched after my first General Magic interview.

I was twenty-five years old and had never really managed anyone, never built a team. Now I was one of the CTOs in a massive company of almost 300,000 people. I'd experienced plenty of failure, but this was truly a new and exciting set of experiences to fail at. The rush of imposter syndrome was almost overwhelming.

Then they told me that anyone who joined the team would have to be drug tested.

Nothing clears your mind like something that unbelievably stupid. Nobody in Silicon Valley would put up with that—pee into a cup to get an engineering job? I wouldn't be able to hire a soul. So I told Philips, "Fuck no!" The fuck was silent but written all over my face. Then I made a deal: "I'll take your test. And if I test negative, no one else I hire on our team will have to take it." Luckily, I tested negative and we hired some amazing people.

Then we got to work negotiating with General Magic to get a version of their operating system (OS) that would do what we needed. I knew the code, knew we could make it work. But by that point General Magic was sinking fast. No revenue, no customers, lots of panic. Marc Porat had made many promises to many people and they were all coming up empty. After months of struggling to squeeze an OS out of General Magic, I got the call: Tony, we just can't do it. Sorry!

I was left with a job title, a budding team, a budget, and a mission we believed in, but no operating system and half a year down the drain. So we gave up the dream of saving General Magic, reluctantly picked Windows Microsoft CE as our OS, and got to work.

If General Magic was a blank slate, a hundred artisans lovingly crafting each element from scratch, then Philips was a Lego set. Here are all the pieces we have. Go make something.

And we did. In 1997 we launched the Philips Velo.

Fig
2.0.1

Released in August 1997, the Philips Velo was 6.7 x 3.7 inches, weighed 374 grams, and cost $599.99. It allowed mobile professionals to email, work on spreadsheets and docs, and update their calendar. Velo's software was built on Windows CE, but its hardware guts were from General Magic.

Velo looked a lot like my pitch at General Magic: a touchscreen and a keyboard, a simpler interface, an explicit focus on business tools.

The next year came the Philips Nino, Velo's smaller sibling.

Velo and Nino both won awards, got critical acclaim. They were the fastest, best-feeling, and had the best battery life of any Windows CE device of the time. I can confidently say that we made the best tools for a Windows-focused businessperson on the go.

We launched a whole marketing campaign, did TV and print ads, and waited for the customers to roll in.

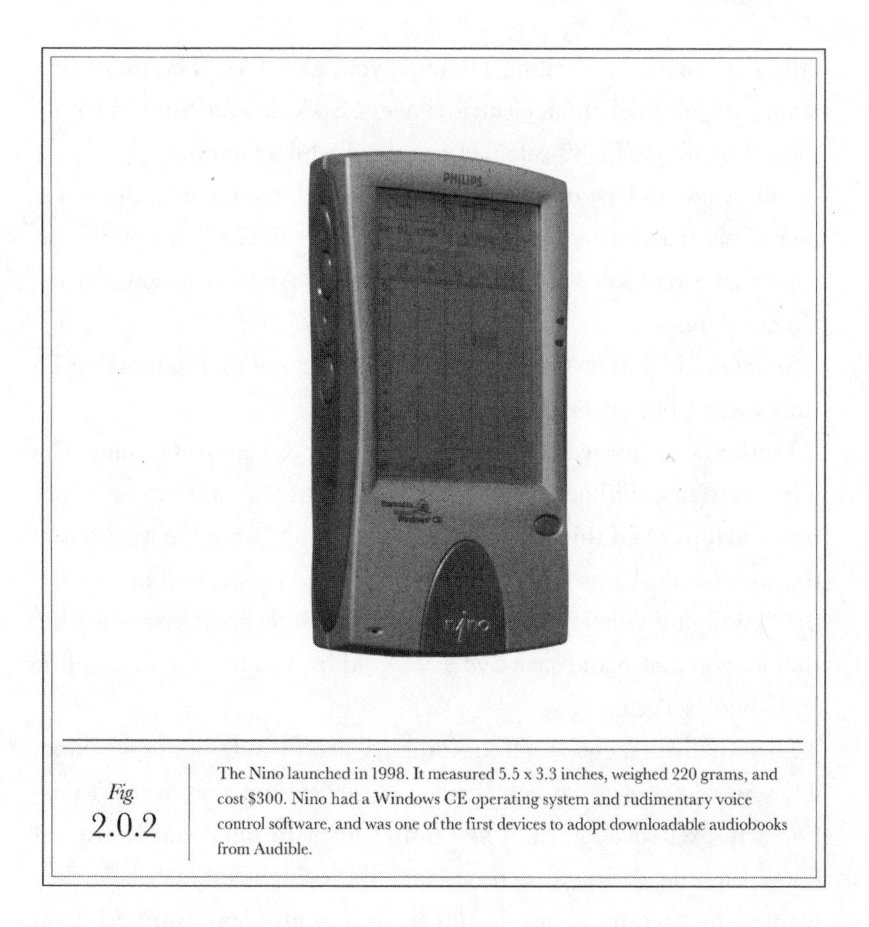

Fig. 2.0.2 The Nino launched in 1998. It measured 5.5 x 3.3 inches, weighed 220 grams, and cost $300. Nino had a Windows CE operating system and rudimentary voice control software, and was one of the first devices to adopt downloadable audiobooks from Audible.

But back then electronics were sold in brick-and-mortar stores and could only fit into two categories: TV/audio or computer equipment. There was no "new technology" aisle. There were printers on one side of the building and stereos on the other and best of luck knowing where a PDA might fit.

So Best Buy put the Velo in the Calculator section.

Circuit City put the Nino in with Laptops.

And customers had no idea where to find them. When they asked the sales reps, they'd get blank stares.

Nobody had any idea how to sell our products. Where to sell them. Who to sell them to. Not the retailers. Not Philips. Our sales team only got bonuses for selling DVD players and TVs. The marketing team was thinking about electric shavers. So Velo and Nino ended up shoved behind a TI-89 calculator and a Toshiba laptop.

Sales were understandably not great, though not terrible. But it was incredibly frustrating—we had put together all the right pieces except one: a real sales and retail partnership. Another lesson learned via gut punch.

So it was time to do something different. But not so different that I'd walk away from my employment contract.

I shifted to a new team: the Philips Strategy & Ventures Group. The job was to help Philips build out its digital strategy and invest in hot new startups. And this is when hot new startups were bursting out of the woodwork. I was a kid in a candy store. We invested in TiVo—the first digital video recorder, a revolutionary technology at the time that let you pause and save live TV—and in Audible, the first online audiobook service.

I'd actually first encountered Audible when I was building the Nino. They were about to launch their own device, but they weren't particularly excited about it. They didn't want to build hardware but knew they needed it to demonstrate the content marketplace they wanted to become. They would have happily demonstrated it on someone else's hardware—but there was nobody else making devices that could play audio. Not even for their tiny single-channel mono spoken-word files.

So Nino became one of the first devices in the world to adopt Audible. It took off. People loved it.

And if we could play audiobooks, why not music? We'd just need a bigger memory. Stereo. Better sound output quality.

I spent a lot of time thinking about it, playing around with the technology. In 1999 I sent out an invitation to my thirtieth birthday party—it was a mixed custom CD I'd burned with full Red Book audio as well as MP3s—"Gimme Some Lovin'," "Instant Karma," "Private Idaho." Even though almost nobody owned an MP3 player back then.

But I could see the potential for a new kind of device: one designed purely for audio.

I talked about it for three hours one day with the CEO of RealNetworks, an incredibly popular technology at the time, the first to create internet streaming audio and video. I'd set up a meeting between the CEOs of our companies to get RealNetworks software on Philips hardware. But the Philips CEO was late. Really late.

By the time he showed up, I had a new job.

I joined RealNetworks to build them a new kind of music player. They told me I could create a team in Silicon Valley and use their technology to form a new vision. Their recruiter was very convincing and, truthfully, the best thing I found at Real. When I met various team leaders, I realized some were ridiculously political. I mean that literally—one is in the Senate now. They tried to get me to sign lengthy noncompetes. And on my first day, they went back on their promise and told me I'd have to move to Seattle. I stepped into my new, tiny, hidden office, ducked around the giant structural pole in the middle of it, and gave my notice after two weeks.

It wasn't easy. The choice to stay or go, to collect a paycheck or save your sanity, to stick with the big company or jump ship to your own venture—it's difficult for everyone.

Just like the specter of management. How do you manage a team when you've never managed before? How do you make decisions when everyone's split on what to do? How do you set a process to make forward progress toward a unified goal? How do you know if you're headed in the right direction? Or if you should quit?

The sooner you realize these questions exist, the better. Everyone rising in their career has to face them at some point.

And I'll be honest: the first time you encounter them, you'll probably screw up. Everyone does. That's okay. You'll learn and grow and get better. But to make that first big leap into leadership less daunting, I've written up a few things that might help.

Chapter

2.1

JUST MANAGING

If you're thinking of becoming a manager, there are six things you should know:

1. **You do not have to be a manager to be successful.** Many people assume that the only path to more money and stature is managing a team. However, there are alternatives that will enable you to get a similar paycheck, have similar amounts of influence, and possibly be happier overall. Of course if you want to be a manager because you think you'll love it, then absolutely pursue it. But even then, remember that you don't have to be a manager forever. I've seen plenty of people go back to being individual contributors, then turn around and be managers again in their next job.

2. **Remember that once you become a manager, you'll stop doing the thing that made you successful in the first place.** You'll no longer be doing the things you do really well—instead you'll be digging into how others do them, helping them improve. Your job will now be communication, communication, communication, recruiting, hiring and firing, setting budgets, reviews, one-on-one meetings (1:1s), meetings with your team and other teams and leadership, representing your team in those meetings, setting goals and keeping people on track, conflict resolution, helping to find creative solutions to intractable problems, blocking and tackling political BS, mentoring your team, and asking "how can I help you?" all the time.

3. **Becoming a manager is a discipline.** Management is a learned skill, not a talent. You're not born with it. You'll need to learn a whole slew of new communication skills and educate yourself with websites, podcasts, books, classes, or help from mentors and other experienced managers.

4. **Being exacting and expecting great work is not micromanagement.** Your job is to make sure the team produces high-quality work. It only turns into micromanagement when you dictate the step-by-step process by which they create that work rather than focusing on the output.

5. **Honesty is more important than style.** Everyone has a style—loud, quiet, emotional, analytical, excited, reserved. You can be successful with any style as long as you never shy away from respectfully telling the team the uncomfortable, hard truth that needs to be said.

6. **Don't worry that your team will outshine you.** In fact, it's your goal. You should always be training someone on your team to do your job. The better they are, the easier it is for you to move up and even start managing managers.

••

You're great at what you do. An amazing accountant, for example. And your team wants a manager who deeply understands their work, who can help them and represent them to leadership. So you work hard to get promoted and you get the job. Congratulations—now you lead an accounting team.

No problem. You're an accountant who's going to tell other accountants how to do their jobs, right? You can do that. This is going to be an amazing team.

So you get deep into the weeds, deep into everyone's work. And they're doing all kinds of weird stuff—not how you would do it at all. And why is it taking so long? Doing their job well is how you got promoted, so now you'll just show everyone how to do it right. You'll tell them—step by step, detail by detail—how to be successful.

It doesn't go well. The team doesn't feel like you trust them. And since you're stretched so thin, focusing on every detail of everyone else's process, nobody really knows what they should be working on or what's most important. People start complaining to you and about you. Everyone gets pissed off.

The more things go wrong, the more you fall back on doing what you know. And you know accounting. So instead of becoming a better accounting manager, you focus on being the best accountant on the team. You start taking on more tasks that your team should be doing themselves. You withhold your feedback and concerns because you don't want to demoralize people further. You rally the team by crying, "We'll get through this! I'll show you how it's done! Just watch and follow me!"

And that's all it takes. That's how normal, reasonable people turn into unbearable micromanagers. That's how projects slow down and disintegrate for lack of leadership. That's the hole that many people fall into when they're put in charge of a team. And some never climb out of it.

Because once you're a manager, you're no longer an accountant. Or a designer. Or a fisherman. Or an artist. Or whatever it is you really enjoyed doing. I constantly have to remind people: If you're doing what you loved in your old job, then you're probably doing the wrong thing. You now lead a team of people doing what you used to be good at. So at least 85 percent of your time should be spent managing. If it's not, then you aren't doing it right. Managing is the job. And managing is hard.

When I was CTO at Philips, my team put up one of those red flashing lights in the office—the old ones that used to be on police cars. They'd turn it on when there was a problem or if they thought I was in a bad mood. They mysteriously knew when some person or group was going to be called into my office soon to have a little talk. Sometimes a passionately loud little talk.

The police light was a joke. Kind of.

I had a team of around eighty people. I was a VP and CTO. I was twenty-five years old. And I was a first-time manager.

I'd never gone through any management training. I'd never even had a real manager myself. I had no mental model of a good manager that I could emulate.

Although my startups had employees, there was no real organization structure. No top-down process, no performance reviews, no meetings to clarify roles and responsibilities. I was a founder, but not a real CEO. Mostly I was an individual contributor on a team of 5–10 people—so we were all just in it together. Nobody was managing anybody. [See also: Figure 5.2.1, in Chapter 5.2]

It was kind of like that at General Magic, too. Our culture was clear: we didn't need managers. Everyone was smart and could manage themselves. So anyone who tried to be a real manager was pretty much ignored.

It was great. Until the team grew. Until we needed to launch something and turn all these brilliant minds in one direction. Until we all had to agree on what was necessary and what got cut.

So when I joined Philips after General Magic, I knew my team would need more structure, that there would have to be defined deadlines and a plan and clear leadership. I knew I'd have to be a manager.

No problem, I thought. I'm an engineer who's going to tell other engineers how to do their jobs, right?

Enter the police light. The stress. And the frustration, for me and the team, the incessant questions and prodding. The micromanagement.

When you're a manager, you're no longer just responsible for the work. You're responsible for human beings. And while that seems obvious—yes, that's the whole point of the job—it's a difficult thing to grapple with when all of a sudden eighty people are looking at you, expecting you to know how to lead them.

So before you decide to be a manager, you should think hard about whether it's the right path for you. Because you don't have to do it. Especially if you don't really want to, but believe the management ladder is the only way to move up in your career. A lot of people shouldn't be forced into management—if you're not really a people person, or you only want to focus on the work, or you thrive on having regular day-to-day successes and accomplishments and the murky maybe-your-team-will-succeed-one-day style of management is less motivating to you.

A star individual contributor (IC) is incredibly valuable. Valuable enough that many companies will pay them just as much as they'd pay a manager. A truly great IC will be a leader in their chosen function and also become an informal cultural leader, someone who people across the company will seek out for advice and mentorship. Apple formally recognizes and rewards star IC engineers in the Distinguished Engineer, Scientist or Technologist (DEST) program. Google Level 8 engineers have similar amounts of clout and sway. Recognition of incredible ICs is most common in engineering, but it's becoming more common in other disciplines.

When considering this path, just make sure you get a very clear view of where you can go in a company over time as an IC. Larger orgs often have clearly defined levels—so find out what the trajectory of an IC is in that organization to understand if they'll value your work.

Many companies also offer the option of being a team lead—or at least they should. This is a kind of a midpoint between an IC and manager. You have some authority to critique, shape, and drive the team's output, but nobody reports to you and you're not dealing with budgets, org charts, or management meetings.

That could have been my path. I could have stayed an engineer. Maybe a team lead. It certainly would have been simpler. Quieter.

But when I finally started looking around at General Magic, I realized coding and designing hardware wasn't as interesting to me as seeing how the whole product—the whole business—came together. [See also: Chapter 1.4: Don't (Only) Look Down.] It became exceedingly obvious that I could never guarantee success through great engineering alone. The best technology wouldn't always win—look at Windows 95 versus Mac OS.

There were all these other squishy elements that needed to align to give any project any chance. Sales, marketing, product management, PR, partnerships, finance—they were all foreign and mysterious and hugely, sometimes desperately, necessary. While I was staring at my feet, workingworkingworking, making the most of a $5 million engineering budget, marketing was getting $10–15 million. I needed to understand why. So I asked.

And that's what changed everything. As soon as I started talking to different teams, I realized my superpower.

A lot of engineers only trust other engineers. Just like finance people only trust finance people. People like people who think like them. So engineers often keep their distance from sales, marketing, creative—all the functions that are soft, squishy.

It's just like how many marketing, sales, and creative teams often don't talk to engineering. Too many numbers. Too black and white. Too many geeks in one room geeking out.

But I wanted to understand the squishy stuff and the geeky stuff. And I liked all of it. I could also translate back and forth—explain the squish to engineers, translate the 1s and 0s to the creatives. I could synthesize all the pieces and keep the whole company in my head.

For me, that was thrilling, exciting, inspiring. It was all I wanted to do. And that meant being a manager. I was drawn to the work, but more importantly, the mission required it. The team needed it.

So I learned how to back off—at least a little bit.

One of the hardest parts of management is letting go. Not doing the work yourself. You have to temper your fear that becoming more hands-off will cause the product to suffer or the project to fail. You have to trust your team—give them breathing room to be creative and opportunities to shine.

But you can't overdo it—you can't create so much space that you lose track of what's going on or are surprised by what the product becomes. You can't let it slide into mediocrity because you're worried about seeming overbearing. Even if your hands aren't on the product, they should still be on the wheel.

Examining the product in great detail and caring deeply about the quality of what your team is producing is not micromanagement. That's exactly what you should be doing. I remember Steve Jobs bringing out a jeweler's loupe and looking at individual pixels on a screen to make sure the user interface graphics were properly drawn. He showed the same level of attention to every piece of hardware, every word on the packaging. That's how we learned the level of detail that was expected at Apple. And that's what we started to expect of ourselves.

As a manager, you should be focused on making sure the team is producing the best possible product. The outcome is your business. How the team reaches that outcome is the team's business. When you get deep into the team's *process* of doing work rather than the actual work that results from it, that's when you dive headfirst into micromanagement. (Of course sometimes it turns out that the process is flawed and leads to bad outcomes. In that case, the manager should feel free to dive in and revise the process. That's the manager's job, too.)

It helps to agree on the process early. To define it up front—

here's our product development process, here's our design process, our marketing process, our sales process. Here's our schedule and how we work and how we work together. Everyone—manager and team—signs off on it and then the manager has to let go. They let the team work.

Then they make sure everything's heading in the right direction in regular team meetings.

These meetings should be structured to get you and the team as much clarity as possible. You should have a weekly crib sheet that helps you keep your priorities and the questions you need to ask top of mind. [See also: Chapter 4.5: Killing Yourself for Work: <u>Everyone thought I was crazy</u>.] Write down a list of what you're worried about for each project and person so you can immediately see when the list is getting too long and you need to either dive deeper or back off.

The other place where you'll get useful data is in 1:1s with team members. It's all too easy to turn 1:1s into friendly chats that go nowhere, so just as you need to have a process for your team meetings, your weekly meetings with individuals should have an agenda, a clear purpose, and should be beneficial to both sides. You should get the info you need about product development and your team members should get insight into how they're doing. Try to see the situation from their point of view—talk about their fears and your own concerns out loud, reframe your thoughts so they can hear the feedback, understand the goals, clear up ambiguities or concerns.

And don't be scared of admitting that you don't know all the answers. You can say, "Help me." If you're a first-time manager or simply new to the company or group, just tell people.

"I'm doing this for the first time. I'm still learning. Please tell me what I can do to make things better."

That's it. But that's a huge mindset shift. I've seen way too many people clam up—terrified that everyone will realize they don't know what they're doing. But of course you don't know what you're doing—pretending you do will fool no one and only dig yourself in deeper. If you've been promoted to management for the first time, you're probably managing people who used to be your peers. Peers who know and trust you. So hold on to that trust. Tell them, "I know I'm your manager now, but we can still talk just like we always have."

And then just be honest with them. Even if things aren't going well, don't avoid telling them the hard truth. Tear off the Band-Aid. If either of you is nervous, you can start the conversation with something positive, ease into it, but don't ignore the elephant in the room, don't tiptoe around the reason you need to talk. It's important to remember that even if you have to criticize someone's work or their behavior, you're not doing it to hurt them. You're there to help. Every word should come from a place of caring. So tell them what's holding them back. Then make a plan to work on it together.

You'll most likely be giving formal written performance reviews every six months—or potentially more often if you're at a company like Google or Facebook where it seems like you're constantly in some kind of review cycle. But those formal reviews should simply be an exercise in writing down the things you're talking about every week. The team should be getting your feedback—good and bad—in the moment rather than waiting to be surprised by it a few months down the road.

I wish there were some magical formula I could tell you for how I figured all this out, but aside from trial and error, I mostly worked on improving myself. I didn't become an engineer by just snapping my fingers and getting a job—I worked at it. I went to school. I practiced for years. The same process is necessary for management.*

I started with some management classes. No class will give you

* If you'd like to hear more about this, I dig into management and my journey through it in the *Tim Ferriss Show* podcast.

all the answers, but any class is better than nothing. And then I went way beyond the basic classes you'll take at a big company—I went down the rabbit hole. I started reading management books and realized that a great deal of management comes down to how you manage your own fears and anxieties. That led me to psychology books. And that led me to therapy. And yoga. I started both in 1995, long before either was widely accepted. It wasn't because I was a crazy person or because becoming a manager was turning me into one. I did therapy and yoga for the same reasons: to find balance, to change the way I reacted to the world, to better understand myself and my emotions and how others perceived them.

The key for me was separating the problems of the company from my personal issues, identifying when my own actions were causing frustrations on the team versus knowing that some things were entirely out of my control. That is difficult stuff to figure out by yourself—it's hard to dig around in your own brain, just like it's hard to do yoga at first without an instructor. My therapist was my coach, my teacher—he helped me understand why I was being such a micromanager. He showed me which parts of my personality I needed to control to effectively lead a team.

Before I learned to create a little distance between what I felt and what I needed to express at work, I let too many of my worries and fears leak into my voice and into my daily interactions. Your team amplifies your mood, so when I was frustrated, those feelings rocketed around the office and came back tenfold. The more upset I got with our lack of progress, the more those frustrations infected the rest of the team. So I had to learn to modulate myself. To crank my personal style down a couple of notches to establish an effective management style.

But I didn't try to change who I was. You are who you are. If you have to completely rearrange your personality to become a manager, then it will always be an act and you won't get comfortable in the role.

I am a loud and passionate person. I will never be Sundar Pichai, CEO of Google and Alphabet. Sundar is quiet, gentle, brilliant, and highly analytical—he always slowly thinks things through before giving a measured response. I pretty much only have one volume setting: slightly loud with a fast crescendo to VERY EXCITED. My son gave me a decibel meter as a gift once—a joke, sure, but it turns out I'm regularly at 70–80 decibels. I am a noisy restaurant, an alarm clock, a vacuum cleaner. Even if every business book on Earth told me to be a quiet, calming voice on the team, I couldn't do it all the time.

So my leadership style is loud and passionate, mission-focused above all else. I pick a goal then run full speed ahead, refusing to let anything stop me, and expect everyone to run with me.

But I also realize that what motivates me may not be what motivates my team. The world is not made up entirely of Tony Fadells (and let us all be grateful for that). There are also normal, sane people with lives and families and lots of things they can and need to do, all pulling at their time.

So as a manager, you have to find what connects with your team. How can you share your passion with them, motivate them?

The answer, as usual, comes down to communication. You have to tell the team why. Why am I this passionate? Why is this mission meaningful? Why is this small detail so important that I'm flipping out right now when nobody else seems to think it matters? Nobody wants to follow someone who throws themselves at windmills for no reason. To get people to join you, to truly become a team, to fill them with the same energy and drive that's bubbling within you, you need to tell them the why.

And sometimes you need to supplement it with a little what. What am I getting paid? What will I be rewarded with if I succeed? Even if your team is all fired up about the mission, don't forget about extrinsic motivation. These are human beings. They may need a raise, a promotion, or even a party. A kind word. Figure out what makes them feel valued. Understand what makes them happy at work.

Helping people succeed is your job as a manager. It's your responsibility to make sure they can become the best versions of themselves. You need to create a setting where they can surprise you. And where they can surpass you.

A lot of people resist that idea—they don't want to hire people to do their jobs, and hiring someone who can do your job better than you is even more terrifying. It's something I hear from new startup CEOs over and over: "Well, if I hire someone to do that . . . then what am I going to do?!"

And the answer, of course, is that whatever you're hiring for isn't your job anymore. If you're a manager or leader or CEO, then your job is to be a manager or leader or CEO. You need to let go of taking pride in your individual daily accomplishments and start taking pride in the accrued wins of your team.

Kwon Oh-hyun, the former CEO of Samsung Semiconductor and an incredible partner, big brother, and sometimes mentor to me as we worked together closely on the iPod, once put it this way: "Most managers are afraid that the people who work for them are going to be better than them. But you need to think of being a manager more like being a mentor or a parent. What loving parent wants their child NOT to succeed? You want your kids to be more successful than you, right?"

Of course there's a natural anxiety about getting shown up. The thought usually goes: "Wait, how can I manage Jane if Jane is better than me? If she excels at this and I don't, then everyone will think she should be doing my job."

And I'm here to break it to you—that might be true. And that's a good thing.

Because if someone under you does something spectacular, that just shows the company that you've built a great team. And that you should be rewarded for it. There should always be at least one or two people on your team who are natural successors to you. Those are the people you have more 1:1s with, who you pull into leadership meetings, who everyone will begin to notice.

The more they notice them, the better. That will make it much easier for you to get promoted, because there will be no question about who can run your team when you move into another role.

There's a reason everyone congratulates parents when their kids do something great—because the kid's achievement is their own, but it also reflects the parent's influence. The parent can take pride in their kid's accomplishment because they know all the time, effort, guidance, hard conversations, and hard work that went into it.

If you're a manager—congratulations, you're now a parent. Not because you should treat your employees like children, but because it's now your responsibility to help them work through failure and find success. And to be thrilled when they do.

One of the people I managed at Apple was Matt Rogers. He was the first intern on the iPod engineering team while he was still in college. Five years later he was senior manager of iPod and iPhone software. He was an obvious superstar—an incredible person and an incredible talent. When I left Apple and started thinking about starting another company, I met with Matt. We became cofounders and together created Nest.

At Nest we hired an intern named Harry Tannenbaum. Harry is analytical, indefatigable, strategic. Five years later, he was director of business analytics and e-commerce at Google Nest. A year after that, he was director of hardware at Google. After Matt left Nest, he called Harry. They started their company in 2020.

I am so damn proud of both of them.

And I can't wait to meet the next generation of talent who they find and mentor and start companies with.

If you're a good manager and build a good team, that team will blast off. So lean into it. Cheer them on when they get promoted. Glow with pride when they kick ass at a board meeting or present their work to the entire company. That's how you become a good manager. That's how you start to love the job.

Chapter

2.2

DATA VERSUS OPINION

You make hundreds of tiny decisions every day, but then there are the critical ones, the ones where you're trying to predict the future, the ones that will put a lot of resources on the line. In those instances, it's important to realize what kind of decision you're faced with:

Data-driven: You can acquire, study, and debate facts and numbers that will allow you to be fairly confident in your choice. These decisions are relatively easy to make and defend and most people on the team can agree on the answer.

Opinion-driven: You have to follow your gut and your vision for what you want to do, without the benefit of sufficient data to guide you or back you up. These decisions are always hard and always questioned—after all, everyone has an opinion.

Every decision has elements of data and opinion, but they are ultimately driven by one or the other. Sometimes you have to double down on the data; other times you have to look at all the data and then trust your gut. And trusting your gut is incredibly scary. Many people don't have either a good gut instinct to follow or the faith in themselves to follow it. It takes time to develop that trust. So they try to turn an opinion-driven business decision into a data-driven one. But data can't solve an opinion-based problem. So no matter how much data you get, it will always be inconclusive. This leads to analysis paralysis—death by overthinking.

If you don't have enough data to make a decision, you'll need insights to inform your opinion. Insights can be key learnings about your customers or your market or your product space—something substantial that gives you an intuitive feeling for what you should do. You can also get outside input: talk to the experts and confer with your team. You won't reach consensus, but hopefully you'll be able to form a gut instinct. Listen to it and take responsibility for what comes next.

••

At General Magic we talked constantly about building a product for Joe Sixpack, but nobody had ever met the guy.

We did user testing as we finished up engineering, but I'm pretty sure we did little to no user research before that. We had no idea what Joe Sixpack might want, so we built features that we liked and just assumed the rest of the world would fall in line.

I was an individual contributor then. I figured leadership knew what they were doing. [See also: Chapter 1.4: Don't (Only) Look Down.]

Then I went to Philips. Now I was leadership. And the pendulum swung hard.

No more assumptions. No more building by intuition. I'd brought a bunch of General Magic people with me and we were all recovering from the face-melting failure of the Magic Link. We knew we couldn't make the same mistakes again. We had to understand our target customer and exactly what they wanted. This time our product was going to be based on clear-cut, black-and-white data. And in the nineties, that meant consumer panels. They were all the rage.

So we hired an external consulting firm and told them we were targeting "mobile professionals." They set up panels in different states, paying thirty to forty people a hundred bucks each to come see our presentation for a few hours.

Then we showed them everything. *Everything.*

At one point we had ten different prototypes for the tiny little keyboard on the Velo. Which one felt better? Which looked more usable? Which felt more reliable? Did you look at the keyboard or the screen while typing? Did you type with all your fingers? Just your thumbs? Do you like the gray? The black? The blue? The bluish gray?

We pored over the videotapes of the sessions. Watched their faces, watched their fingers, studied their answers on our little forms. Then the consultants would do the same—they'd collate everything and give a report six weeks later.

The customer is always right, right?

Except customer panels can't design for shit. People just can't articulate what they want clearly enough to definitely point in one direction or another, especially if they're considering something completely new that they've never used before. Customers will always be more comfortable with what exists already, even if it's terrible.

But we fell into the same trap as everyone else. We were wowed by the consultants, excited by the numbers. And we quickly became far too reliant on them: everyone wanted data so they wouldn't have to make decisions themselves. Instead of moving forward with a design, you'd hear, "Well, let's just test it." Nobody wanted to take responsibility for what they were making.

So you'd run the test. And then run it again. On Monday the customer panel would pick option X. On Friday, the same group would go with option Y. Meanwhile, we were paying millions of dollars to consultants who took a month and a half to put their own slant on everything.

The data wasn't a guide. At best, it was a crutch. At worst, cement shoes. It was analysis paralysis.

And this doesn't just happen with old-school customer panels. If it had been 2016 rather than 1996, we would have leaned into A/B testing—the omnipresent tool of the internet age. A/B testing just means running a digital experiment where you test option A versus option B with customers. So some see a blue button, some see an orange button, and you see which button gets the most clicks. It's an incredible tool—infinitely faster than customer panels and so much easier to interpret.

But even with A/B testing, we probably would have gotten the same muddled results, the same product-killing fear of making the wrong decision.

Despite the fact that many companies now rabidly test every single element of their product and unquestioningly follow the clicks, A/B and user testing is not product design. It's a tool. A test. At best, a diagnosis. It can tell you something's not working, but it won't tell you how to fix it. Or it can show you an option that solves one hyperlocal issue but breaks something else downstream.

So you have to design the options and the tests to really know what you're testing. You have to think through what A and B are rather than let them be randomly assigned by an algorithm or thoughtlessly thrown against a wall to see what sticks. And that takes insight and knowledge of the entire customer journey. You need a hypothesis, and that hypothesis should be part of a bigger product vision. So you can A/B test where the "Buy" button should go on a Web page, whether it should be blue or orange, but you shouldn't be testing whether or not a customer should buy online.

If you're testing the core of your product, if the basic functionality can flex and change depending on the whims of an A/B test, then there is no core. There's a hole where your product vision should be and you're just shoveling data into the void.

In our case—and in the case of every first-generation product—we could have been shoveling forever. There was never going to be enough data to make an assured choice.

If a product is really new, there's nothing to compare it to, nothing to optimize, nothing to test.

We were right to define our target customer clearly, to talk to them and find out what problems they had. But then it was our job to figure out the best way to fix those problems. We were right to ask their opinions and get feedback about our designs. But then it was our job to use those insights to move forward in a direction we believed in.

Eventually our team figured it out—we stopped throwing money at consultants, stopped spinning in circles, started moving forward and trusting ourselves and the trusted opinions of smart people around us.

We made decisions. I made decisions. This is in. This is out. This is how it's going to work.

Not everyone on the team agreed with me. That'll happen sometimes when one person has to make the final call. In those moments it's your responsibility as a manager or a leader to explain that this isn't a democracy, that this is an opinion-driven decision and you're not going to reach the right choice by consensus. But this also isn't a dictatorship. You can't give orders without explaining yourself.

So tell the team your thought process. Walk through all the data you looked at, all the insights you gathered, and why you ultimately made this choice. Take people's input. Listen, don't react. There may be a minority of team members who agree with the decision; there might be some good feedback that makes you modify your plan. If not, give the speech: I understand your position. Here are the points that make sense for our customers, here are the ones that don't. We have to keep moving and, in this instance, I have to follow my gut. Let's go.

Even if some people on your team don't love that answer, they'll respect it. And they'll trust you—they'll know that they can speak up and criticize your choices and not get immediately shot down. And then they can sigh, and shrug, and go back to their team, communicate the "why" of the decision, and get on the train.

That's what's always worked for me. It's how my team at Philips came to accept my decisions.

However, the ever-changing leadership at Philips never did. Right up until launch, they were asking us for data to prove that a market for our products existed. But when you're making something new, there's no way to definitively prove that people will like it. You just have to ship it—put it out into the world (or at least in front of forgiving customers or internal users) and see what happens.

It's important at this stage to have a boss who understands the kinds of decisions you're facing. You need a leader who trusts you, who's ready to back you up.

But those kinds of leaders—those kinds of human beings—are hard to find.

Most people don't even want to acknowledge that there are opinion-driven decisions or that they have to make them. Because if you follow your gut and your gut is wrong, then there's nowhere else to cast blame. But if all you did was follow the data and you still failed, then clearly something else was wrong. Someone else screwed up.

This is often a tactic of people who are trying to cover their asses. *It's not my fault! I just went where the data sent me! The data doesn't lie!*

That's why some managers and execs and shareholders demand data even when there is none and then chase that imaginary data directly into the abyss. These are the kinds of people who won't question their directions and drive their car right off a cliff. If at all possible they want to erase the human element—human judgment—from the equation.

These are also the people who will call in the big-league, very expensive (and in my opinion worthless) consultants at the drop of a hat. They'll happily second-guess your decision, then rip it right out of your hands and pass it to people who have no context or understanding of your product, company, or culture.

When that happens, you need to figure out what's going on so you can try to steer management in a different direction. Here are a few reasons why a leader may sit on your idea and then call in the consultants:

1. **Delay.** They may be waiting for something—a promotion or a bonus—and don't want to take a risk until they get it.
2. **Fear for their job.** They may be convinced that the consequence of failure is that they'll lose this project or their position or—if the failure is spectacular enough—their job.
3. **They don't have the time or don't want to bother.** They don't believe it's worth the effort to dig in and really understand the decision, choose from the array of options before them, and take a risk. They just want someone else to do it and make them look smart.
4. **They know what they want but don't want to hurt anyone's feelings.** They want to be seen as "nice" so they'll just keep testing the water, asking for more data again and again until you're worn out and exasperated.

So what do you do when you're stuck with a manager who's hell-bent on driving off a cliff, ideally while throwing all their money out the window at some consultants? Or what if you have data but it's inconclusive—nobody can say for sure where it leads? Or what if you need to convince your team to follow you even though you can't prove you're heading in the right direction?

You tell a story. [See also: Chapter 3.2: Why Storytelling.]

Storytelling is how you get people to take a leap of faith to do something new. It's what all our big choices ultimately come down to—believing a story we tell ourselves or that someone else tells us. Creating a believable narrative that everyone can latch on to is critical to moving forward and making hard choices. It's all that marketing comes down to. It's the heart of sales.

And right now you're selling—your vision, your gut, your opinion.

So don't just hit them with the classic "This is Jane, this is her life, and this is how her life changes when she uses our product" slide. Helping people see things from the customer's perspective is a critical tool, but it's just part of what you need to do. Your job in this moment is to craft a narrative that convinces leadership that your gut is trustworthy, that you've found all the data that could be gleaned, that you have a track record of good decisions, that you grasp the decision makers' fears and are mitigating those risks, that you truly understand your customers and their needs and—most importantly—that what you're proposing will have a positive impact on the business. If you tell that story well, if you bring people along with you on that journey, then they will follow your vision, even if there's no hard data to back you up.

Nothing in the world is ever 100 percent sure. Even scientific research with entirely data-based outcomes is actually filled with caveats—we didn't do this kind of sampling, there was this variant, we need to follow up with this test. The answer may not be the answer. There's always a chance we're wrong.

So you can't wait for perfect data. It doesn't exist. You just have to take that first step into the unknown. Combine everything you've learned and take your best guess at what's going to happen next. That's what life is. Most decisions we make are data-informed, but they're not data-made.

As the brilliant, empathetic, refreshingly insightful, and egoless designer Ivy Ross, vice president of hardware design at Google, has said, "It's not data or intuition; it's data and intuition."

You need both. You use both. And sometimes the data can only take you so far. In those moments, all you can do is take a leap. Just don't look down.

Chapter

2.3

ASSHOLES

Throughout your career, you'll encounter some real assholes. These are (mostly) men and (sometimes) women who come in different flavors of selfish or deceitful or cruel, but have one unifying characteristic: you cannot trust them. They can and will screw you and your team over, either to get something for themselves or just to push you down and make themselves look like the hero. You'll find them at all levels, as individual contributors and managers, but the greatest concentration is near the top—according to University of San Diego professor Simon Croom, up to 12 percent of corporate senior leadership exhibit psychopathic traits. [See also: Chapter 5.1: Hiring: Everyone on the team knew what we interviewed for.]

However, you will also encounter people who can be very difficult to work with—who are gruff or loud or bossy or infuriating—and who may, at first, seem like assholes, but whose motivations and actions tell a very different story.

It's important to realize what kind of person you're dealing with so you can understand how best to work with them—or, if necessary, how best to get around them.

Here are the different assholes you might have to deal with:

1. **Political assholes:** The people who master the art of corporate politics, but then do nothing but take credit for everyone else's work. These assholes are typically incredibly risk averse—they're

focused exclusively on surviving and pushing others down so they can reach the top. They don't make anything themselves—are absent for the real work and tough decisions—but they'll happily leap in to cry "I told you so" when anybody else's project has a hiccup, then try to swoop in to "fix" it. They often won't speak up in large meetings because they never want to be seen by their bosses as being wrong—they can't risk looking like an idiot. Instead they'll work in the background to undermine you and everyone else who isn't on their "team." These assholes usually build a coalition of budding assholes around them—copycats who see it as their path to success. And there's always one person who they hate and plot against and have to push out of the way somehow.

2. **Controlling assholes:** Micromanagers who systematically strangle the creativity and joy out of their team. These assholes can never be reasoned with. They resent any good idea that didn't come from them and are extremely threatened by anyone on their team who is more talented than they are. They never give people credit for their work, never praise it, and often steal it. These are the assholes who dominate big meetings—who won't let you get a word in edgewise, and who get defensive and angry if anyone critiques their ideas or suggests alternatives. These assholes are sometimes really good at what they do—they hone their skills to a fine point, then use it to cut down everyone around them.

3. **Asshole assholes:** They suck at work and everything else. These are the mean, jealous, insecure jerks who you'd avoid at a party, but who inevitably sit immediately next to you at the office. They cannot deliver, are deeply unproductive, so they do everything possible to deflect attention away from themselves. They will lie, craft gossip, and manipulate others to get people off their scent. The only good thing about these assholes is that they're generally out the door pretty quickly—they can only deflect for so long before people start

noticing that they bring zero value. And nobody likes working with them.

In addition, assholes can act in different ways:

Aggressive: They freak out. They yell. They accuse you of all kinds of nonsense. They sneer at you in a meeting and demean you in front of your manager. These assholes are easy to spot.

Passive-aggressive: They smile. They nod. They agree with you, act friendly. Then they go behind your back, spread vicious gossip, and try to screw you at every step. This is by far the more dangerous variety of asshole—you don't see them coming until you feel the knife in your back.

You will also encounter another kind of person at work who often gets confused with the controlling asshole. While the knee-jerk reaction is to dismiss them as just another egotistical jerk, they have a very different motivation: it's always to make the work better, not to benefit themselves or hurt others. Most significant, you can trust them. They may not always make decisions you like, but they're focused on the greater good and will listen to reason if it's in the best interest of the product and customer. This makes them fundamentally different from true assholes. But it doesn't make them any easier to work with:

4. **Mission-driven "assholes":** The people who are crazy passionate—and a little crazy. They speak most frankly, trampling the politics of the modern office, and steamroll right over the delicate social order of "how things are done around here." Much like true assholes, they are neither easygoing nor easy to work with. Unlike true assholes, they care. They give a damn. They listen. They work incredibly hard and push their team to be better—often against their

will. They are unrelenting when they know they're right, but are open to changing their minds and will praise other people's efforts if they're genuinely great. A good way to know if you're working with a mission-driven "asshole" is to listen to the mythos around them—there are always a few choice stories floating around about some crazy thing they've done, and the people who've worked with them closely are always telling everyone that they're not that bad, really. Most tellingly, the team ultimately trusts them, respects what they do, and looks back at the experience of working with them fondly, because they pushed the team to do the best work of their lives.

••

Plenty of people think I'm an asshole.

It's usually because I get loud. I ask nicely a few times and then—if we're still not getting anywhere—I stop asking nicely. I put pressure on myself and the people around me. I don't let up. I expect the best—from myself, from everyone else. I care deeply about our mission, our team, our customers. I can't stop myself from caring.

So I push. If something seems off, if I think there's a chance we can do better, that the customer can get more, then I do not let up. I do not let things slide. [See also: Chapter 6.1: Being CEO.] I push people who are experts, who already know how it's done, how it's always been done, to find a new way to do it. And that's a lot to take. It's not easy to work with. I would never claim otherwise.

But pushing for greatness doesn't make you an asshole. Not tolerating mediocrity doesn't make you an asshole. Challenging assumptions doesn't make you an asshole. Before dismissing someone as "just an asshole," you need to understand their motivations.

There's a world of difference between being emphatic and passionate to benefit the customer versus bullying someone to appease your own ego.

It's not always an obvious difference to the person on the other end. It's hard to walk into a hurricane and think, Ah, this is a passionate hurricane. I just need to let it blow for a little while and then present some helpful data.

But some hurricanes can be reasoned with. Some cannot.

So here's how to deal with people like me, how to talk down a hurricane: ask why.

It's the responsibility of a passionate person—especially a leader—to describe their decision and make sure you can see it through their eyes. If they can tell you why they're so passionate about something, then you can piece together their thought process and either jump on board or point out potential issues.

So ask. Don't be afraid to push. They'll respect you more if you stand up for what you believe in. Mission-driven "assholes" want to be better at their jobs and fulfill that all-important mission—they want to make sure the company is heading in the right direction.

So if it's in the best interest of the customer, they'll hear you and change their mind. Eventually.

It's something I was always telling my team at Apple whenever Steve Jobs went completely off the rails: "Yes, this idea is insane. But sanity will prevail! Even if Steve is wrong today, trust that he'll get to the right answer sooner or later. We just need to find a better approach and make our case."

Prepare for some wind and some hail, but don't worry about getting swept away: a mission-driven "asshole" might tear apart your work, but they won't attack you personally. They won't call you names or fire you for disagreeing with them.

That's the difference between a mission-driven "asshole" and a controlling one.

Controlling assholes won't listen. They'll never admit they screwed up. Neither will political assholes. They'll ignore obvious problems and deflect reasonable feedback, either because it's not helpful politically or because their ego can't take it. They don't protect the product or the customer or the team. They protect themselves.

And to be clear, Steve Jobs was not one of those assholes. Of course he crossed the line sometimes—he was human—but I didn't condone it or excuse it and it wasn't the norm. Steve was a mission-driven "asshole," a passionate hurricane.

The best thing for the product would always win out eventually because the product was all that mattered. Steve was always focused on the work. Always.

It's the assholes who are focused on people—on controlling people—who make work miserable. Real assholes always make it personal. Their motivation is their ego, not the work. As long as they're winning, they don't give a shit about what's happening to the product or what the customer has to deal with. These are the assholes who make it progressively more difficult to create something you're proud of.

Like the manager who told a friend of mine flat out, "Do not talk to the CEO!"

During product development the CEO would often call her with questions or ideas or just to brainstorm. He couldn't get the information as fast as he wanted from her manager so he'd come straight to her.

Her manager was livid. How could she flout the pecking order? That's not how we do things around here!

So he said, "Never talk to the CEO. Never call him. Never email him. Only go through me."

But she wasn't calling him. He was calling her. And she wasn't dumb. If the CEO wanted to talk, she was going to answer. She offered to tell her manager everything they discussed, but that wasn't good enough. Instead of leaning into the work so he'd have the answers the CEO wanted, he just forbade her from speaking.

She rolled her eyes and completely ignored his order. But she had to deal with this guy to get her project shipped. So she did the only thing you can do when faced with a controlling asshole:

1. Kill 'em with kindness.
2. Ignore them.
3. Try to get around them.
4. Quit.

In that order.

Start by giving them the benefit of the doubt. Maybe they've just had bad experiences before or they had a crappy relationship with whoever they used to work with on your team. Maybe they just don't understand how to work with you. Maybe this is all just a giant misunderstanding—you'll get over it and show them this can be a productive relationship.

First make sure you're not the problem—that there's not something you did to give the wrong impression or accidentally cause an issue. Have an open conversation where you acknowledge that you got off on the wrong foot. Be friendly. Be kind. Try to praise them publicly—give them credit for something they've done (even if they did it wrong). Sometimes that's all it takes.

Sometimes it isn't.

After you've given it your very best shot—consulted with them, gotten their advice, treated them fairly, had some honest conversations—and you've gotten a big pile of nothing in return, then you go on the defensive. If you have a good manager, ask them to protect you from the asshole. See if they can rearrange things so you don't have to deal with this person and no longer have to hear their input.

If that doesn't help, then just ignore them. Stop involving them in your decisions. Ask for forgiveness, not permission—and eventually don't even bother asking for that. If you're doing something valuable to the company and clearly worthwhile, then the asshole can yell or scheme all they want, but their hands will mostly be tied. Don't be aggressive or unpleasant about it—just do your thing.

Sometimes that can buy you enough time to complete your project in peace.

Sometimes it can't.

After weeks of being ignored, after trying to cut me down in meeting after meeting, tiny slight after tiny slight, one asshole I worked with pulled me into his office—with HR present—looked me straight in the eye, and said, "There are two swinging dicks in this room. And mine is the biggest."

I'll give him one thing: that was hard to ignore.

I remember sitting there trying to process that statement. What did he expect me to say? Or do? Did he want me to take a swing at him? Was that the goal? It was such a bizarre moment, such a record scratch, that I did the exact right thing. I sat there in silence and stared at him. He kept going—that was just his opening salvo. But I didn't get into an argument. I didn't engage. I just rearranged my understanding of the world. Okay, so that's who this guy really is. This is the game we're playing. He is not on my team. He doesn't deserve my respect.

Now I needed to go on the offensive. And I needed backup.

If you're having trouble with an asshole, then typically you won't be the only one exasperated. So find people who agree with you that this asshole has to go—talk to their peers, talk to HR. Find the right moment and talk to their boss—they'll usually give you a nod and say they're already doing something about it. It'll probably take forever and be very messy, but hopefully they'll either get off your project or entirely out of your life.

If that doesn't work, you can try to transfer teams. But when you're dealing with a real asshole, their reputation is probably well known in the company. If another team knows that taking you on will bring on the ire of the asshole in question, they might decide it's not worth the hassle. I remember one instance where a person became a pariah—no other team wanted them, for fear that the losing manager would seek revenge.

At that point, your only option might be the very last one on the table.

Quit.

Tell your boss and HR and whoever else is paying attention that you've tried everything and you can no longer work with this person. [See also: Chapter 2.4: I Quit: But when you're at the end of your rope.]

If you're valued and useful, leadership will probably scramble to try to keep you and find a way to diffuse the situation. The key is always to deliver on a meaningful project. If you're delivering and they're not, eventually the assholes are recognized as assholes and become isolated or ineffective. It can take a long, long time, but usually their opinions begin to lose their power and they fade away.

But not always.

Sometimes, even if they're chased out of the organization, they can still screw you over.

So always keep an eye on social media. Don't just watch for internal rumblings; remember to check Glassdoor, Facebook, Twitter, Medium, LinkedIn, hell—even Quora. TikTok. Whatever. Pissed-off people will poison the water anywhere. Social media is a new weapon in every asshole's arsenal. If they fail to get what they want from you at work, they can make things very, very personal very, very publicly.

This is always problematic and incredibly unpleasant, but if they're controlling assholes or just your regular, run-of-the-mill asshole-assholes, they'll probably undermine themselves and the truth will come out eventually.

Political assholes are a whole other animal.

The trouble with political assholes is that they often form coalitions with other political assholes. Otherwise pleasant people will watch the assholes get promoted and think that's the right path for them. So the coalition of assholes will grow and they'll focus almost exclusively on managing up, so leadership won't realize what's happening.

Political assholes thrive in large organizations where they can pull the kind of Machiavellian BS that makes you sound crazy and paranoid when you're describing it. They find people who aren't exceptional at their jobs and protect them in exchange for their allegiance. They get dirt on their peers—who's having an affair with his admin? Can we get HR to cover it up?—then those people are indebted for life.

It's like the mafia. But instead of killing people, they kill good ideas.

Political assholes need an army to sow seeds of discord or get gossip and funnel it up. That's how they control people. That's how they get away with it.

So how do you fight the mafia?

You gather together the people you work with and make a plan to step up your game. But you don't do it to protect yourselves, or to get promotions or power or bonuses or whatever the assholes are after. You band together in service of your customers.

Political cliques are tit-for-tat, *Survivor*-esque pyramids, each asshole scrambling and fighting to be on top. Your group should be focused on raising each other up and protecting customers from the assholes' terrible decisions. When an alliance of assholes starts spreading lies, or stealing ideas, or taking over projects they have no business touching, they will parrot each other's words to leadership. They will make sure they all have the same narrative. They'll back each other up until they're impossible to ignore.

That's when your team needs to have a counternarrative. The bullshit-asymmetry theory, Brandolini's law, will be at play here: "The amount of energy needed to refute bullshit is an order of magnitude higher than to produce it."

So you need to craft a great story and walk into meetings ready to support each other. Agree ahead of time—make sure everyone knows the script. Gather data to back you up so it's not just your word against theirs. Then when the asshole pipes up, your crew will have the ammunition and manpower to call them out.

Hopefully you'll be able to neutralize the mafia, or at least get them to focus their efforts on easier prey. And the one good thing that comes out of these kinds of battles is that you'll forge lasting connections with a diverse group of wonderful people.

After we stopped the assholes from ruining the product and screwing over customers, we could stop crafting narratives. Stop playing the stupid games we never wanted to play in the first place. We could get back to the work we loved.

That's the thing about assholes—they're so incredibly unpleasant that they stand out in your memory. They get a whole chapter in your book. But most people just want to go to the office and make something cool. The vast majority of people who cause you trouble aren't malicious or Machiavellian—they're struggling, or first-time managers, or in the wrong job, or just having a really, really bad day. Maybe their kid's not sleeping. Maybe their mom died. Even the nicest people on Earth can act like assholes sometimes. Or maybe they're passionate hurricanes who are pushing you further than you thought you could go, because they know you're talented and that you're holding yourself back.

Most people aren't assholes.

And even if they are, they're also human. So don't walk into a job trying to get anyone fired. Start with kindness. Try to make peace. Assume the best.

. . . and if that doesn't work, then remember that what goes around comes around. Although it never comes fast enough.

Chapter

2.4

I QUIT

Stick-to-it-iveness is an important value. If you're passionate about making something, you'll need to doggedly pursue it, and that may mean earning less money for a while or staying at a problematic company so you can finish your project.

However, sometimes you just need to quit. Here's how you know:

1. **You're no longer passionate about the mission.** If you're staying for the paycheck or to get the title you want, but every hour at your desk feels like an eternity, then save yourself. Whatever you're staying for is not worth the soul-sucking misery of a job you cannot bear to get out of bed for.

2. **You've tried everything.** You're still passionate about the mission but the company is letting you down. So you've talked to your manager, to other teams, to HR, and to senior leadership. You've tried to understand the roadblocks and pitched solutions and options. But still, your project is going nowhere or your manager is impossible or the company is falling apart. In that case, you should leave that job but stick to the mission and find another team on a similar journey.

Once you do decide to quit, make sure you leave in the right way. You've made a commitment, so follow through and try to finish as much of what you started as possible. Find a natural breakpoint in your project—the next big milestone—and aim to leave then. The longer you're at a company, and the higher up you are, the longer it will take to transition out. Individual contributors can usually give a few weeks to a couple months of notice. CEOs may need a year or more.

••

I quit Philips after seeing my projects through—ensuring I had explored every avenue to make my team successful. I quit because we were never going to outshine the competition when everyone was using the same Microsoft operating system that dictated most of our features. I quit after four years of hard work and frustration and learning and personal and professional growth.

I quit RealNetworks after two weeks because I could see the writing on the wall: I was going to hate that job.

Even so, I stayed an extra four weeks after I gave my notice. I wrote up options for different businesses they could start, sketched out business plans and project presentations. I wanted to make sure I left them with something tangible—real work built around good ideas—so nobody could say he came and left and screwed us (although I'm sure they said it anyway).

But I needed to get out of there. The second they went back on their word and told me to move to Seattle, I lost all trust in that company. And you cannot work with people you cannot trust. Everything inside of me screamed that this was only going to go from bad to worse.

Most people know in their gut when they should quit and then spend months—or years—talking themselves out of it. But I could tell from the start that I would have been well paid and utterly miserable.

And I want to make it very clear: hating your job is never worth the money.

I need to repeat that: hating your job is never worth whatever raise, title, or perks they throw at you to stay.

I know that may ring hollow coming from me, a lucky, wealthy person. But the way I've gotten wealthy is not by accepting giant paychecks or titles to do jobs I know I'll hate. I follow my curiosity and my passion. Always. And that's meant leaving money on the table—so much money that people thought I might actually be crazy. "Look at what you are walking away from—leading iPhone, leaving Apple? And all that money? What's wrong with you?"

But it's been worth every cent.

Anyone who's ever stuck with a job they hated knows the feeling. Every meeting, every pointless project, every hour stretches on and on. You don't respect your manager, you roll your eyes at the mission, you stagger out the door at the end of the day exhausted, dragging yourself home to complain to family and friends until they're as miserable as you are. It is time and energy and health and joy that disappear from your life forever. But hey, that title, that stature, that money—it's worth it all, right?

Don't get trapped. Just because you don't know of any other better options doesn't mean they don't exist. There is other money. There are other jobs.

Once you put out word that you're looking or that you've left your job, new opportunities will most likely come to you. I see this happen with friends all the time. They post an update on LinkedIn and people immediately reach out. Oh! This person is available. That's exciting.

Of course, as with everything, it helps to know the right people.

The key to finding them is networking. By that I don't mean going to a conference and working the room, handing out your business cards or QR codes and cornering potential employers as they try to eat their stack of tiny sandwiches. I just mean make new relationships, beyond business—talk to people outside your bubble. Get to know what else is out there. Meet some new human beings. Networking is something you should be doing constantly—even when you're happily employed.

I remember getting lunch in 2011 with an exec who had just left Apple and was going to start a new company. He'd worked at Apple since the late nineties and had been a protégé of Steve Jobs for years before that. And you'd think he'd have every advantage in the world—he'd spent the last decade working at the highest levels in the heart of Silicon Valley's most famous company, by the side of its most dynamic leader. Who wouldn't fund him? Who wouldn't jump at the chance to work with him?

But it was as if he'd just gotten out of prison. He'd never talked to anyone outside of Steve's sphere of influence. He didn't know who to go to, how to raise money. His only relationship to the world was through Apple, and once he left he was clueless. He figured it out, of course, eventually. But it took so much longer than he expected.

So don't get trapped.

And don't think of networking as a means to an end—as a tit-for-tat exchange where if you do someone a favor they may do you one in return. Nobody wants to feel like they're being used.

You should talk to people and make connections because you're naturally curious. You want to know how other teams at your company work and what people do. You want to talk to your competitors because you're all working to solve the same problems and they're taking a different approach. You want your projects to be successful, so you don't just talk to your immediate teammates at lunch—you grab lunch with your partners, your customers, their customers, their partners. You talk to everyone: get their ideas and their perspectives. In doing so you may be able to help someone or make a friend or strike up an interesting conversation.

And an interesting conversation can turn into an interview. Or not. But at least it'll be interesting. At least you might feel the spark of potential. And that may send you down another pathway, with another conversation. And another. And another. Until you see a light at the other end: a company or job or team that makes you want to go to work again. That helps you start feeling like yourself again.

And when that happens, quit your old job. Quit, quit, quit.

But don't just walk into your manager's office and throw your notice on their desk and walk away from everything you've worked on. Even if you hate your job, don't leave it in a tangle of loose ends. Finish what you can, clean up what you can't, and hopefully transition it to the next person who's inheriting your responsibilities. It may be weeks, or even months. If you're a manager or senior leader, it'll honestly feel like forever—I had a nine-month transition out of Google Nest. At Apple, it took twenty months.

People won't remember how you started. They'll remember how you left.

But don't let that deter you from making the choice and getting out.

Once you find yourself in a place where you believe in the mission, everything changes.

Of course, you may need to quit that job, too. Because once you're committed to a mission, to an idea—that's the thing you should stick to. The company is secondary. If you find something that inspires you, then follow the best opportunities to pursue it. I got hooked on personal electronics and followed that passion across five companies. It only became really lucrative at the very end, but it was what I loved to do, so I kept finding new opportunities to do it. Each job took a different angle, a new perspective on the same problem, and eventually I had a rich, 360-degree view of the challenge I wanted to solve and all the possible solutions. The idea was much more precious than the company signing my paychecks.

But it's a balance. With RealNetworks it was just bad juju all around, an immediate loss of trust, but the other companies I worked at for four years, five years, almost a decade. If you've found a good opportunity to follow your passions, you should not give up until you've tried to make it work at the company you're at.

So if something's not working, don't just complain to people who have no power to fix it, then throw up your hands and quit. Even talking to your manager isn't enough. Especially if it's your manager that's the problem.

If the mission you're excited about is growing dimmer because of internal politics or poor administration or leadership churn or simply bad decisions, don't be shy. Get to networking. Talk to everyone. Not watercooler talk or internal gossip, not just complaining with no solutions. Come with suggestions to fix the intractable problems that you and your team face. Speak to your manager, HR, other teams—find appropriate leaders who will listen. Hopefully some will agree with you, or challenge your view or help you refine your thinking. It's all useful. Get their perspectives.

That includes senior leadership. Executives. You could even go to board members and investors if you can get ahold of them. That's what I did—at Philips, at Apple. Get up as high as you can and let them know what the issues are. You'll probably quit anyway if these issues aren't solved, so you have nothing to lose.

Most people at the top are interested to hear what's happening down below. They may reward you for bringing it to their attention. They may even share your frustrations (although they might not tell you that).

And yes, you'll probably drive your boss nuts. Going around your manager is always touchy. I drove my managers completely insane every time I sidestepped them to reach out to some other executive. So if they ask, tell your boss what you're doing—and tell them why. This is a time to ask for forgiveness, not permission. Explain that you've talked to them (and you should have first) but nothing is getting fixed. Tell them what you're worried about and your proposed solutions; explain who you're reaching out to and what you hope to accomplish.

But if you take this route—if you go around your boss and start making a fuss all over the company—make sure the issues you're raising are not about yourself.

I remember we had a huge all-hands meeting at Apple once—these meetings would only happen two, maybe three times a year. And a guy stands up during the Q&A and starts asking Steve Jobs why he didn't get a raise or a good review. Steve looks at him in stunned disbelief and says, "I can tell you why. Because you're asking this question in front of ten thousand people."

He was fired shortly thereafter.

So don't be that guy.

You can have personal problems—not enough money, glacial career progression—or you can have issues with the project you're working on. Quitting because of personal problems is completely valid; complaining about them to everyone at the company is not. And you don't have to mess up in front of ten thousand people, either—whining constantly to just one executive about your stock grants is almost as bad.

If you're going to get everyone's attention, make sure it's to support the mission, not for personal gain. Think through the problems that are plaguing your project. Write down thoughtful, insightful solutions. Present them to leadership. Those solutions may not work, but the process will be at the very least educational. Don't nag, but be persistent, choose your moment wisely, be professional, and don't hold back about the consequences if you don't succeed. Tell them you're passionate about making this job work, but if you can't solve these issues then you'll probably quit.

But you have to mean it. It can't just be a negotiation ploy. Too many people trash their career at a company because they had a tantrum. You absolutely cannot threaten to quit, then hesitate and flip-flop and stay. Everyone will instantly lose respect for you. You have to follow through.

The threat of leaving may be enough to push your company to get serious and make whatever change you're asking for. But it might not. Quitting should never be a negotiating tactic—it should be the very last card you play.

And remember that even if leadership acknowledges that you're right and promises a major shift, it may take a while for anything to change. Or it may never change. But it's worth it to try. Quitting anytime things get tough not only doesn't look great on your résumé, but it also kills any chance you have of making something you're proud of. Good things take time. Big things take longer. If you flit from project to project, company to company, you'll never have the vital experience of starting and finishing something meaningful.

Jobs are not interchangeable. Work is not just a sweater you can take off when things get hot. Too many people jump ship the second they need to dig in and really push through the hard, grinding work of making something real. And when you look at their résumés you can instantly see the pattern.

A two-page résumé can tell the story of a three-hundred-page novel once you know what you're looking for. And too many plots have giant holes in them.

So before you quit, you'd better have a story. A good, credible, and factual one. You'll need to have a rationale for why you left. And you'll need one for why you want to join whatever company you're heading to next. These should be two very different narratives. You'll need them for the interview, but also for yourself—to make sure you've really thought things through. And to make sure you're making the right choice for the next job.

Your story about why you left needs to be honest and fair and your story for your next job needs to be inspiring: this is what I want to learn, this is the kind of team I want to work with, this is part of the mission that truly excites me.

Keep this in mind when the recruiters reach out. Because if you're successful, they will. Knowing when to quit and follow a recruiter is a two-stage process: first you have to know your job is no longer for you, and then you have to decide the new place is better. Too many people conflate the two, get dazzled by the recruiter's sales proposition, and ignore the opportunities they have where they work right now. Or they don't network internally so they don't even realize what opportunities there are. I've seen too many people jump ship before doing their research and really thinking things through. They usually come back three to six months later, tail between their legs, sheepishly asking for their old job back.

So don't be that person, either.

But when you're at the end of your rope, truly at the end—not just impressed by a recruiter—then don't be scared to walk away.

I quit Apple three times. The first time was right after we launched the iPod. Our team moved mountains to deliver it months faster than anyone thought possible, to rave reviews. And we did it despite the fact that my manager was doing his best to take credit for our team's hard work [See also: Chapter 2.3: Assholes.]

I'd tried everything—engaging him, ignoring him, fighting him, soothing his ego—but now the project was done. My team had worked nonstop for ten months. So I asked for what was promised—the title I should have already had. "When do I get VP?"

And he said, "Let's wait a year. These things take time. No one gets promoted that fast."

He knew damn well that I deserved a higher title from the start, that he'd stiffed me on the way in (you can read the full story in Walter Isaacson's *Steve Jobs*, if you're curious). But now I had delivered. I'd overdelivered.

I tried to stay calm. I explained my reasoning. He simply shrugged and gave a half grin. "Sorry. Not now."

The last scrap of respect I had for him flew out the window.

I still believed in the mission. I was proud of what we'd made. I was excited to keep going. But there was no getting around this guy. He was going to screw me over no matter how great a job I was doing. This was a wound that would not heal.

Enough was enough. So I said the only thing that was left to say: "I quit."

Sometimes the only way to save yourself is to walk out the door.

Two weeks later, as I was packing up my office, I got a call from Cheryl Smith, the human resources leader charged with overseeing our iPod team. She was an incredible partner who'd opened my eyes to how the Apple machine worked and helped me navigate it when I was a newbie. "I heard what happened," she said. "It makes absolutely no sense. You can't leave! Let's go for a walk."

The longer we walked around the Apple campus and I told her the details of what went down, the louder we got, the more we gestured wildly in the air. She was empathetic, said she'd work on it, told me to hold tight—but I figured it was too late. Twenty-four hours later I was going to leave Apple for good.

The next day, a few hours before they were planning to escort me out, I got a call from Steve Jobs.

"You're not going anywhere. We'll get you what you want."

I marched over to my manager's office. Cheryl was waiting outside with a wide grin.

My manager had reluctantly come to the table, although he was grimacing, hating every minute of it. "This is not how we do things around here," he muttered as he signed the paperwork to get me my promotion.

That evening, I walked into my goodbye party and said, "I'm staying!"

As time passed I had to quit again. This time to protect the product and the team. And then once more—that time to protect my sanity and my family. And there was drama, of course. A tremendous amount. Turning away from my team, from Steve, was not easy.

But I knew it was the right move. After a decade of dedicating all my energy to Apple, it was time to walk away.

Sometimes all the calculations, negotiations, discussions with your manager and meetings with HR are entirely beside the point. Sometimes it's just time to go. And when that moment comes, you'll probably know.

Quit and go do something you'll love.

Part

III

BUILD YOUR PRODUCT

The basic technology for the first iPod wasn't designed at Apple.

It wasn't even designed for a handheld device.

In the late nineties people began filling up their hard drives with MP3 audio files. For the first time high (enough) quality music could be stored in small enough files that you could download vast song libraries to your computer.

But even if you had a lavish stereo system to listen to that music on, you couldn't use it. Stereos were built for tapes and CDs, so everyone just played their newly downloaded music out of their crappy computer speakers.

In 1999 I saw the potential for something better. Not an MP3 player—a digital audio jukebox.

It would let you convert all your CDs to MP3s so you could listen to them, plus whatever you downloaded, on your TV and home stereo system. Before the iPod's famous tagline "1000 songs in your pocket," we were trying to make "1000 CDs in your home theater."

That's what I pitched to RealNetworks, anyway. But that was the wrong place, wrong people, wrong everything. So I figured—screw it. I'll do it myself.

The words that launched a thousand startups.

I called mine Fuse Systems.

The inspiration came from a project at Philips. They'd tried to build a home theater + DVD player that ran Windows so you could browse the internet on your TV and sort of stream audio from the Web (as well as you could stream anything pre-Wi-Fi).

It was the kernel of a good idea. Home internet connections were speeding up, from 56kbps to a blazing 1mbps, making audio and even grainy, stamp-sized video downloads possible. It became clear that people's music and movie collections were going to move onto computers. But nobody wanted to listen to music on the sad, gray, corporate Windows computers we had in the nineties. Home theaters were much better—they had HDTV and surround sound. But only the most refined audiovisual geeks could install them.

Philips could see that, but they couldn't capitalize on it. They got elbow deep in Microsoft, building a PC with delusions of being a stereo. They were focused on what they could make, not why anyone would want it. I looked it over and thought: No. Nonono. You can't use Windows—I'd been bashing my skull against a Microsoft OS for years and knew it was a dead end for consumer electronics. Who wants to wait two minutes for their TV to boot up? And you had to simplify the home theater for nongeeks. Make something that anyone could just plug and play.

I wanted to build one component that would hook up to the internet, but wouldn't look or feel like a computer. Fuse was going to give people a consumer electronics experience: you'd be able to configure and order a whole home theater, including a CD/DVD player that would save your music to a built-in hard drive. Then you'd connect to the world's first online store to download more songs, and one day, movies and TV shows. TiVo was all the rage then but I wanted Fuse to go further.

I got a little seed money and then I was in it. I had to build a com-

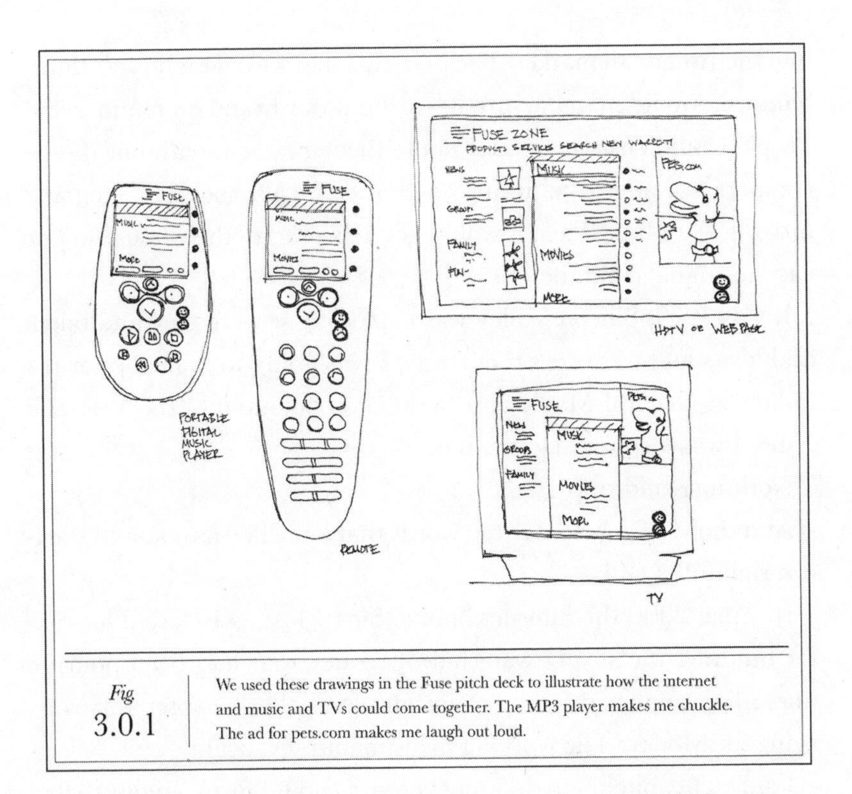

Fig. 3.0.1 We used these drawings in the Fuse pitch deck to illustrate how the internet and music and TVs could come together. The MP3 player makes me chuckle. The ad for pets.com makes me laugh out loud.

pany. And not a side-job, small-potatoes college startup. A real company. A serious business.

I was going to get it right this time. We were going after the world's biggest players. We were going to challenge Sony.

But first I had to convince people to come work with me. I'd walked away from Philips's enormous infrastructure, their mountains of process and cash, and walked right onto a blank slate. I had a big idea but not much else. And all these people I was trying to recruit—they expected to get paid. They expected health care. They expected HR and accounts payable and all the stuff you take for granted when you work at an actual business.

So I got to work. And work. And work. And work.

I built the team—hired twelve people. We partnered with Samsung, at the time a little-known Korean consumer brand trying to break into the American market. The plan was that we'd design everything, Samsung would manufacture it; we'd put our brand on it and sell it. People would customize their home theater system with our digital components, add in Samsung's rebranded TVs, speakers, etc., and order it all online. Then it would get delivered to their house in one easy-to-manage bundle.

It was 1999. Silicon Valley was exploding with money and talent and ideas and we were on our way. I was going to make up for the failure of General Magic and the wasted potential of the Velo and Nino. I was inspired. Determined.

Nothing could stop us.

And those, of course, are the words that launched a thousand startups right off a cliff.

In April 2000 the internet bubble burst. Just as I started looking for funding, the steady waterfall of money that had been pouring into Silicon Valley dried up overnight. [See also: Chapter 4.3: Marrying for Money: The world of investment is cyclical.]

I did eighty pitches to different venture capital firms. Eighty. All of them failed. Investors were scrambling to save the startups they'd already (over)invested in, and nobody was interested in funding pricey consumer electronics when the stock market was tanking, companies were going belly-up, and billions of dollars were flowing down the drain. Timing is everything and my timing couldn't have been worse. I couldn't raise a dime.

One day at the peak of my desperate, scrabbling attempt to fund my company, I had lunch with an old friend from General Magic. I told him what I was working on and what I was struggling with—the swirling, nauseating mix of excitement about what we were creating and the sinking horror that I'd have to shut it all down. He commiserated, ate his sandwich, and wished me well.

The following afternoon he had lunch with a colleague who worked at Apple. They mentioned they were kicking off a new project. Did he happen to know anyone with experience building handheld devices?

I got a call from Apple the next day.

Because you've picked up this book, the rest of the story is probably pretty familiar. I took a consulting gig with Apple at first, just hoping to make enough money to pay my employees or maybe leverage my job into a buyout for Fuse. Putting my hopes on Apple was a serious long shot. Steve Jobs was back at the helm then, but during the previous decade Apple had been in a death spiral, launching a slew of mediocre products that edged the company close to collapse. The Macintosh was struggling to break 2 percent market share in the United States; their computer sales were stagnating. At the time Apple's market cap was around $4 billion. Microsoft's was $250 billion.

Apple was dying. But Fuse was dying faster.

So I took the job.

Fig 3.0.2 | This is the Styrofoam model I made in March 2001 to convince Steve to green-light the iPod project.

- The call from Apple came the first week of January 2001.
- A couple of weeks later I became a consultant leading the iPod investigation. But it wasn't the iPod yet. The code name was P68 Dulcimer—and there was no team, no prototypes, no design, nothing.
- In March, Stan Ng and I pitched the idea for the iPod to Steve Jobs.
- The first week of April I became a full-time employee and pulled the Fuse team with me.
- By the end of April, Tony Blevins and I found our manufacturer, Inventec, in Taiwan.
- In May I hired DJ Novotney and Andy Hodge, the first additions to the original Fuse team.
- On October 23, 2001—ten months from when I started—the iPod emerged into the world, our fat plastic-and-stainless-steel baby.

Fig 3.0.3 This was the first iPod, released in October 2001 with the famous tagline "A thousand songs in your pocket." It was 4.02 x 2.43 inches, had a $399 price tag, and was pretty damn close to the original vision model I'd cut seven months before.

I was incredibly lucky to lead the team that made the first eighteen generations of the iPod. Then we got another incredible opportunity—the iPhone. My team created the hardware—the metal and glass that you held in your hand—and the foundational software to run and manufacture the phone. We wrote the software for the touchscreen, the cellular modem, the cell phone, Wi-Fi, Bluetooth, etc. Then we did it again for the second-generation iPhone. And then again for the third.

I blinked and it was 2010.

I spent nine years at Apple. It's the place where I finally grew up. I wasn't just managing a team anymore. I was leading hundreds, thousands of people. It was a profound shift in my career and in who I was.

After a decade of failure, I finally made something—actually two things—that people actually wanted. I finally got it right.

But it didn't feel like success at first. Or even in the end. It was still work, every step of the way.

Apple is where I learned where to draw the line—is it done enough? Is it good enough?

It's where I learned the real meaning of design.

And it's where I learned to organize my brain and my team in the face of intense, grinding, never-ending pressure.

So if you're headed into a new phase of your career, navigating at higher and higher levels, building teams, building relationships, trying to find your footing further and further away from the actual thing you're making but responsible for so much more than before, stressed out beyond belief, then I'm here to share what I've learned.

Chapter

3.1

MAKE THE INTANGIBLE TANGIBLE

People are easily distracted. We're wired to focus our attention on tangible things that we can see and touch to the point that we overlook the importance of intangible experiences and feelings. But when you're creating a new product, regardless of whether it's made of atoms or electrons, for businesses or consumers, the actual thing you're building is only one tiny part of a vast, intangible, overlooked user journey that starts long before a customer ever gets their hands on your product and ends long after.

So don't just make a prototype of your product and think you're done. Prototype as much of the full customer experience as possible. Make the intangible tangible so you can't overlook the less showy but incredibly important parts of the journey. You should be able to map out and visualize exactly how a customer discovers, considers, installs, uses, fixes, and even returns your product. It all matters.

••

When I was a kid I spent a lot of time with my grandfather building stuff—birdhouses, soapbox derby cars. We'd fix up lawn mowers and bikes or work on additions to the house.

It felt good. So much of life as a child was confusing and out of my control, but there was no ambiguity to physical objects. You built them, held them in your hands, handed them to others. Satisfying. Clean.

Even after I dove headfirst into programming, I didn't question my innate belief that the computer itself was the key to everything. Electrons were nothing without atoms.

That's why I was so excited about joining General Magic after college. I'd been programming and programming but now I was going to make a *thing*. A device, a physical object, a computer like the one that had changed my life.

But the longer I made things—at General Magic, at Philips, at Apple—the more I realized that many things don't need to be made.

After the iPod a lot of people started pitching me their devices. People would say, "Tony's the hardware guy—he'll love your idea." And the first thing I did when someone proudly handed me their beautifully polished prototype was put it aside. "How can you solve your problem without this?"

They would be astonished. How could the "hardware guy" not want to check out my cool gadget?

People often get excited about making something with atoms—they dig into the design, interface, colors, materials, textures—and instantly become blind to simpler, easier solutions. But making anything with atoms is incredibly difficult—it's not an app that you can copy and update with a click. The only time hardware is worth the headache of manufacturing and packaging and shipping is if it's critically necessary and transformative. If hardware doesn't absolutely need to exist to enable the overall experience, then it should not exist.

Of course, sometimes you do need hardware—it can't be avoided. But when that happens, I still tell people to put it away. I say, "Don't tell me what's so special about this object. Tell me what's different about the customer journey."

Your product isn't only your product.

It's the whole user experience—a chain that begins when someone learns about your brand for the first time and ends when your product disappears from their life, returned or thrown away, sold to a friend or deleted in a burst of electrons.

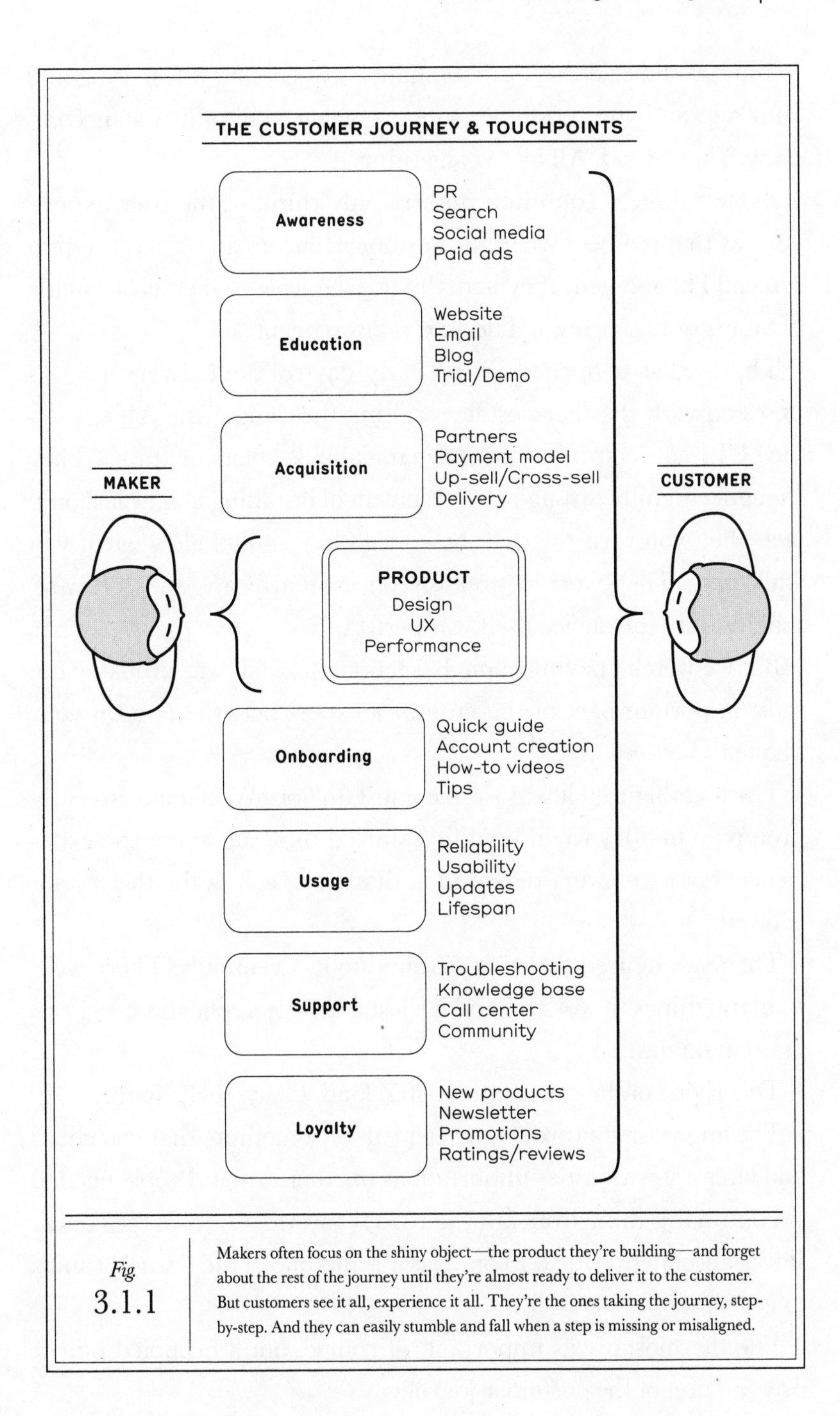

Fig 3.1.1 Makers often focus on the shiny object—the product they're building—and forget about the rest of the journey until they're almost ready to deliver it to the customer. But customers see it all, experience it all. They're the ones taking the journey, step-by-step. And they can easily stumble and fall when a step is missing or misaligned.

Your customer doesn't differentiate between your advertising and your app and your customer support agents—all of it is your company. Your brand. All of it is one thing.

But we forget. Too often makers only think of the user experience as that moment when the customer touches an object or taps a screen. The moment they actually use the *thing*—whether it's made of atoms or bits or both. The *thing* is always central.

That's what happened in the early days of Nest. Everyone was obsessed with the thermostat—crafting the design, the AI, the device UI, the electronics, the mechanical bits, colors, textures. They thought carefully through every element of installing it, how it should feel when you turn the dial, how brightly it should glow when you walk past. They worked tirelessly on the hardware and software, making sure the device itself was perfect.

But we weren't paying enough attention to what was probably the most important part of the customer experience: the app on your phone.

The team figured it was simple—just an app. We'd made an early prototype in 2011 when we'd first started thinking about the experience, but then didn't return to it, didn't revise it as the thermostat evolved.

The team figured they'd get around to it. Eventually. There were so many things to work on and it's just a mobile application. We can figure it out fast.

This is one of the times I got a little loud. Okay, really loud.

The app wasn't a throwaway element or something that you could add later—it was just as important as the thermostat. People needed to control this thing from anywhere in the world. Or from the couch. It was absolutely critical to our success and one of the hardest things to get right.

The thermostat was important, of course, but it occupied only a tiny fraction of the customer journey:

- 10% of our customers' experience was the website, advertising, packaging, and in-store display: first we had to convince people to buy it or at least consider and research it.
- 10% was installation: following the instructions to get it onto your wall with minimal nervousness and power outages.
- 10% was looking at and touching the device: it had to be beautiful so people would want it in their homes. But after a week it learned what you liked and when you were away, so you didn't really need to touch it much. If we did our job right, customers would only interact with it here and there, during unexpected cold snaps or heat waves.
- 70% of the customer experience was on people's phones or laptops: you'd open the app to turn up the heat on the way home, or you'd see how long the AC was on in Energy History, or you'd tweak your schedule. Then you'd check your email and see a summary of how much energy you used that month. And if you had an issue, you'd go to our website and use the online troubleshooter or read a support article.

If we didn't execute well on any one of these parts of the customer experience, Nest would have failed. Each phase of the journey has to be great in order to move customers naturally into the next, to overcome the moments of friction between them.

There are bumps between Awareness and Acquisition, between Onboarding and Usage, between every phase of the journey, that you have to help customers over. In each of these moments, the customer asks "why?"

Why should I care?

Why should I buy it?

Why should I use it?

Why should I stick with it?

Why should I buy the next version?

Your product, marketing, and support have to grease the skids—continually communicate and connect with customers, give them the answers they need, so they feel like they're on a smooth ride, a single continuous, inevitable journey.

To do that right, you have to prototype the whole experience—give every part the weight and reality of a physical object. Regardless of whether your product is made of atoms or bits or both, the process is the same. Draw pictures. Make models. Pin mood boards. Sketch out the bones of the process in rough wireframes. Write imaginary press releases. Create detailed mock-ups that show how a customer would travel from an ad to the website to the app and what information they would see at each touchpoint. Write up the reactions you'd want to get from early adopters, the headlines you'd want to see from reviewers, the feelings you want to evoke in everyone. Make it visible. Physical. Get it out of your head and onto something you can touch. And don't wait until your product is done to get started—map out the whole journey as you map out what your product will do.

That's how you hack your brain. How you hack the brains of everyone on your team.

Start from that very first moment of the customer journey. You should be prototyping your marketing long before you have anything to market.

At Nest, that meant focusing on the box.

The packaging led everything. The product name, the tagline, the top features, their priority order, the main value props—they were literally printed on a cardboard box that we constantly held, looked at, tweaked, revised. The physical limitations of the box forced us to zero in on exactly what we wanted people to understand first, second, third. To fit the tiny space, the creative team crafted crisp descriptions that we could later use in our videos, our advertising, on our website, and in interviews with press. To evoke the Nest brand, they covered the box in warm, rich photos that allowed people to imagine this object in their own homes, their own lives.

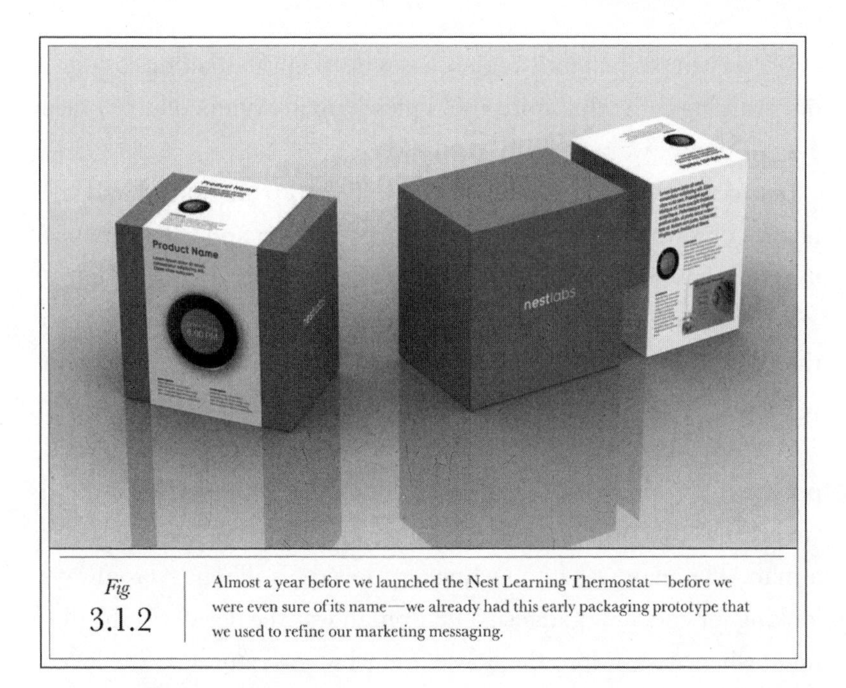

Fig. 3.1.2 — Almost a year before we launched the Nest Learning Thermostat—before we were even sure of its name—we already had this early packaging prototype that we used to refine our marketing messaging.

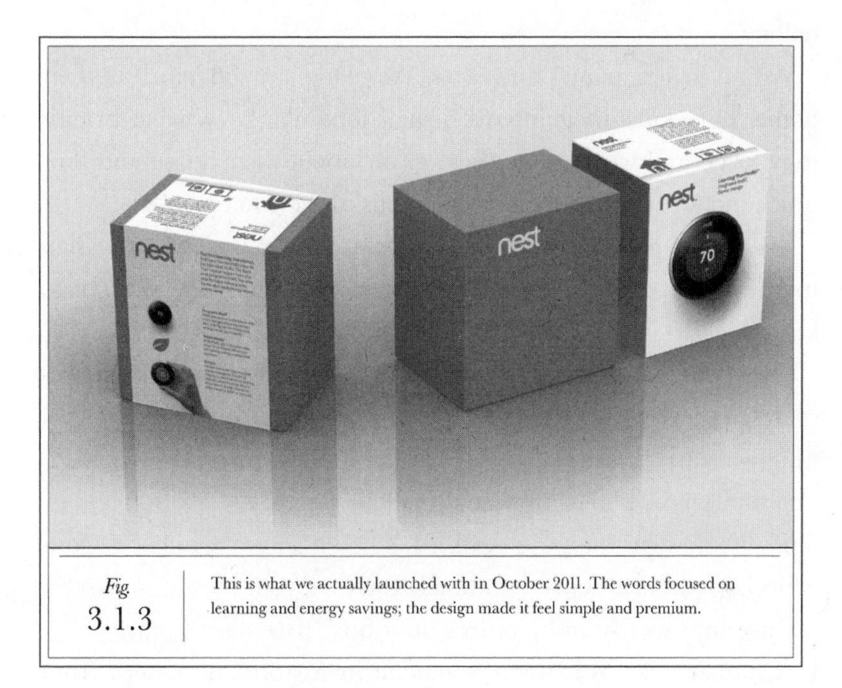

Fig. 3.1.3 — This is what we actually launched with in October 2011. The words focused on learning and energy savings; the design made it feel simple and premium.

We turned the box into a microcosm of all our marketing so someone walking through a store could pick it up and immediately grasp everything we wanted them to know.

But to prototype this moment properly—to truly understand the half second when someone notices the packaging and leans forward to pick it up—you can't just call this theoretical person "someone."

We had to know them. Who were they? Why would they pick up the box? What would they want to know? What was most important to them?

We took everything we'd learned about the industry and Nest's potential customers, about demographics and psychographics, and we created two distinct personas. One was a woman and the other a man. The man was into technology, loved his iPhone, was always looking for cool new gadgets. The woman was the decider—she dictated what made it into the house and what got returned. She loved beautiful things, too, but was skeptical of super-new, untested technology.

We gave them names and faces. We made a mood board of their home, their kids, their interests, their jobs. We knew what brands they loved and what drove them crazy about their house and how much money they spent on heating bills in the winter.

We needed to look through their eyes to understand why the man might pick up the box. And so we could convince the woman to keep it.

Over time we added more personas—couples, families, roommates—as we better understood our customers. But in the beginning we started with two—two human beings who everyone could imagine, whose photos they could touch.

That's how prototyping works. It's how you make abstract concepts into physical representations. You turn your messaging architecture into words and pictures on a box. [See also: Figure 5.4.1, in Chapter 5.4.] You turn "someone in a store" into Beth from Pennsylvania.

And then you keep going. Every step of the way, along every link of the chain.

When we had prototypes of the actual thermostat, we sent it out to real people to test. We knew self-installation was potentially a huge stumbling block, so everyone waited with bated breath to see how it went. Did people shock themselves? Start a fire? Abandon the project halfway through because it was too complicated?

Soon our testers reported in: It went fine. Everything's up and running! But it took about an hour to install.

We winced. Crap. An hour was way too long. Beth from Pennsylvania would not be cool with turning off the power, opening up the wall, and fiddling with unknown wires for an hour. This needed to be an easy DIY project, a quick upgrade.

So we dug into the reports—what was taking so long? What were we missing?

Turns out we weren't missing anything—but our testers were. They spent the first thirty minutes looking for tools—the wire stripper, the flathead screwdriver; no, wait, we need a Phillips. Where did I put that little one again?

Once they got everything they needed, the rest of the installation flew by. Twenty, thirty minutes tops.

I suspect most companies would have sighed with relief. The actual installation took twenty minutes, so that's what they'd tell customers. Great. Problem solved.

But this was going to be the first moment people interacted with our device. Their first experience of Nest. They were buying a $249 thermostat—they were expecting a different kind of experience. And we needed to exceed their expectations. Every minute, from opening the box to reading the instructions to getting it on their wall to turning on the heat for the first time, had to be incredibly smooth. A buttery, warm, joyful experience.

And we knew Beth. Searching for a screwdriver in the kitchen drawer—then the toolbox in the garage; no, wait, maybe it's in the drawer after all—would not make her feel warm and buttery. She would be rolling her eyes by minute five. She would be frustrated and annoyed.

So we changed the prototype. Not the thermostat prototype—the installation prototype. We added one new element: a little screwdriver. It had four different head options, fit in the palm of your hand. It was sleek and cute. But more importantly, it was unbelievably handy.

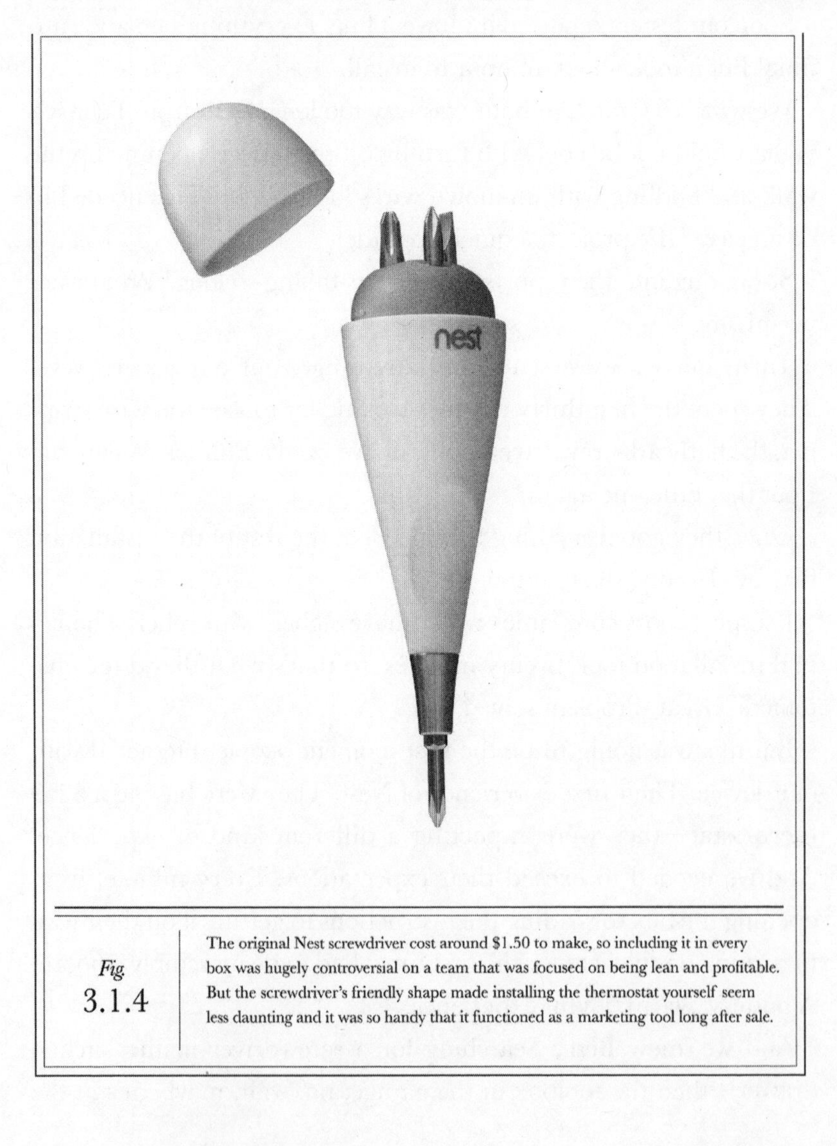

Fig. 3.1.4 The original Nest screwdriver cost around $1.50 to make, so including it in every box was hugely controversial on a team that was focused on being lean and profitable. But the screwdriver's friendly shape made installing the thermostat yourself seem less daunting and it was so handy that it functioned as a marketing tool long after sale.

So now, instead of rummaging through toolboxes and cupboards, trying to find the right tool to pry their weird old thermostat off the wall, customers simply reached into the Nest box and took out exactly what they needed. It turned a moment of frustration into a moment of delight.

And then it turned into a lot more than that.

The screwdriver was never just for installation. It had ripple effects all the way up and down the customer journey.

A vital part of the customer experience is post-sale. How do you stay connected to your customer in a way that's actually useful? How do you keep on delighting people instead of just marketing to them, selling and selling until they're sick of you?

Our thermostat was made to be on people's walls for ten years. By design, it would become like a piece of art—occasionally admired and adjusted, mostly fading into the background.

But every time they opened the random-stuff drawer in their kitchen, they'd see the cute little Nest screwdriver. And they'd smile.

Every time they'd need to replace the batteries in their kid's toy car, they'd grab our screwdriver. And suddenly the screwdriver became the toy and the car was forgotten.

We knew it wasn't just a hardware tool—it was a marketing tool.

It helped customers remember Nest. It helped them fall in love.

And it helped people discover us. Journalists wrote articles about the screwdriver. It appeared in every five-star review. It was free PR, a boost to word of mouth. Instead of a bowl of candy at the Nest front desk, we had a bowl full of screwdrivers. It became a symbol for the entire user experience—thoughtful, elegant, long-lived, and deeply useful.

That's why I wouldn't let anyone strip it out.

It was a constant battle with every new generation of thermostat. The screwdriver was expensive. Each one ate into our margins. So there was always a brigade of employees petitioning to remove it—

they couldn't understand why we'd add to our COGS (cost of goods sold) that way.

But they didn't understand that it wasn't a straight COGS line item. It was a marketing expense. And a support expense. That screwdriver saved us so much money on phone support. Instead of angry calls, we had happy customers raving online about their great experience.

If we hadn't thought through installation with the same care and attention that we lavished on the thermostat, it would never have occurred to us to put a screwdriver in every box.

And if we hadn't thought about the full customer life cycle—from discovery to support to loyalty—we would have just made the kind of tiny, one-use screwdriver that comes with IKEA furniture. Instead we included four heads—more than anyone needed to install the thermostat—so that people could use it for practically anything. So that Nest stayed in their brains as long as the screwdriver stayed in their drawer. Longer.

When a company gives that kind of care and attention to every part of the journey, people notice. Our product was good, but ultimately it was the whole journey that defined our brand. That's what made Nest special. It's what makes Apple special. It's what allows businesses to reach beyond their product and create a connection—not with users and consumers, but with human beings. It's how you create something that people will love.

Chapter

3.2

WHY STORYTELLING

Every product should have a story, a narrative that explains why it needs to exist and how it will solve your customer's problems. A good product story has three elements:

- It appeals to people's rational and emotional sides.
- It takes complicated concepts and makes them simple.
- It reminds people of the problem that's being solved—it focuses on the "why."

That "why" is the most critical part of product development—it has to come first. Once you have a strong answer for why your product is needed, then you can focus on how it works. Just don't forget that anyone encountering your product for the first time won't have the context you have. You can't just hit customers on the head with the "what" before you tell them the "why."

And keep in mind that customers aren't the only ones who will hear this story. Telling the story is how you attract people to your team or investors to your company. It's what your salesperson puts in their slide deck and what you put in your board presentation.

The story of your product, your company, and your vision should drive everything you do.*

---

* If you're interested in more about design and the storytelling behind it, I'd recommend finding my conversation with Peter Flint on his NFX podcast.

••

I remember sitting in the stands watching Steve Jobs tell the world about the iPhone in 2007.

> *This is the day I've been looking forward to for two and a half years.*
>
> *Every once in a while, a revolutionary product comes along that changes everything and Apple has been—well, first of all, one's very fortunate if you get to work on just one of these in your career. Apple has been very fortunate. It's been able to introduce a few of these into the world.*
>
> *In 1984 we introduced the Macintosh. It didn't just change Apple. It changed the whole computer industry.*
>
> *In 2001, we introduced the first iPod. And it didn't just change the way we all listen to music, it changed the entire music industry.*
>
> *Well, today we're introducing three revolutionary products of this class. The first one is a widescreen iPod with touch controls. The second is a revolutionary mobile phone. And the third is a breakthrough internet communications device.*
>
> *So, three things: a widescreen iPod with touch controls; a revolutionary mobile phone; and a breakthrough internet communications device. An iPod, a phone, and an internet communicator. An iPod, a phone . . . are you getting it? These are not three separate devices, this is one device, and we are calling it iPhone. Today, Apple is going to reinvent the phone, and here it is.*

That's the part of the speech that everyone remembers. The buildup and surprise, the brilliant setup. People still write articles about it. They celebrated its ten-year anniversary.

But the rest of the speech was just as important. After the setup, he reminded the audience of the problem Apple was solving for them. "The most advanced phones are called smartphones, so they say. And the problem is that they're not so smart and they're not so easy to use." He talked for a while about regular mobile phones and smartphones and the problems of each before he dove into the features of the new iPhone.

He used a technique I later came to call the virus of doubt. It's a way to get into people's heads, remind them about a daily frustration, get them annoyed about it all over again. If you can infect them with the virus of doubt—"Maybe my experience isn't as good as I thought, maybe it could be better"—then you prime them for your solution. You get them angry about how it works now so they can get excited about a new way of doing things.

Steve was a master of this. Before he told you what a product did, he always took the time to explain why you needed it. And he made it all look so natural, so easy.

I'd watched other CEOs give pitches before, they'd hardly know what their supposedly revolutionary product was. Sometimes they didn't even know how to hold it right. But customers and the press would always be in awe of Steve's presentations. "It's a miracle," they said. "He's so calm, so collected. No prepared speeches, slides with almost no words—he just knows what he's talking about and it all hangs together."

It never felt like a speech. It felt like a conversation. Like a story.

And the reason is simple: Steve didn't just read a script for the presentation. He'd been telling a version of that same story every single day for months and months during development—to us, to his friends, his family. He was constantly working on it, refining it. Every time he'd get a puzzled look or a request for clarification from his unwitting early audience, he'd sand it down, tweak it slightly, until it was perfectly polished.

It was the story of the product. And it drove what we built.

If part of the story didn't work, then part of the product wasn't going to work, either, and would need to be changed. That's ultimately why the iPhone had a glass front face instead of plastic and why it didn't have a hardware keyboard. Because the story of the "Jesus Phone" wouldn't hold together if it got scratched the first time you put it in your pocket or if you'd be forced to watch movies on a tiny little screen. We were telling the story of a phone that would change everything. So that's what we had to build.

And when I say "story," I don't just mean words.

Your product's story is its design, its features, images and videos, quotes from customers, tips from reviewers, conversations with support agents. It's the sum of what people see and feel about this thing that you've created.

And the story doesn't just exist to sell your product. It's there to help you define it, understand it, and understand your customers. It's what you say to investors to convince them to give you money, and to new employees to convince them to join your team, and to partners to convince them to work with you, and to the press to convince them to care. And then, eventually, it's what you tell customers to convince them to want what you're selling.

And it all starts with "why."

Why does this thing need to exist? Why does it matter? Why will people need it? Why will they love it?

To find that "why," you need to understand the core of the problem you're trying to solve, the real issue your customers face on a regular basis. [See also: Chapter 4.1: How to Spot a Great Idea: The best ideas are painkillers, not vitamins.]

And you have to hold on to that "why" even as you build the "what"—the features, the innovation, the answer to all your customers' problems. Because the longer you work on something, the more the "what" takes over—the "why" becomes so obvious, a feeling in your gut, a part of everything you do, that you don't even need to express it anymore. You forget how much it matters.

When you get wrapped up in the "what," you get ahead of people. You think everyone can see what you see. But they don't. They haven't been working on it for weeks, months, years. So you need to pause and clearly articulate the "why" before you can convince anyone to care about the "what."

That's the case no matter what you make—even if you sell B2B payments software. Even if you build deep-tech solutions for customers who don't exist yet. Even if you sell lubricants to a factory that's been buying the same thing for twenty years.

There's a competition for market share and a competition for mind share. If your competitors are telling better stories than you, if they're playing the game and you're not, then it doesn't matter if their product is worse. They will get the attention. To any customers, investors, partners, or talent doing a cursory search, they will appear to be the leaders in the category. The more people talk about them, the greater their mind share, and the more people will talk about them.

So you have to find an opportunity to craft stories that stick with customers and keep them talking about you. Even if your customer knows you and your product, or they're highly technical, there are frictions that you can eliminate for them. You can explain why they need one version of lubricant over the other or give them information they never had before. Or you explain why buying the same product from your company is better than buying that product from a competitor. You can earn their trust by showing that you really know your stuff or understand their needs. Or offer them something useful; connect with them in a new way so they feel assured that they're making the right choice with your company. You tell them a story they can connect with.

A good story is an act of empathy. It recognizes the needs of its audience. And it blends facts and feelings so the customer gets enough of both. First you need enough insights and concrete information that your argument doesn't feel too floaty and insubstantial. It doesn't have to be definitive data, but there has to be enough to feel meaty, to convince people that you're anchored in real facts. But you can overdo it—if your story is only informational, then it's entirely possible that people will agree with you but decide it's not compelling enough to act on just yet. Maybe next month. Maybe next year.

So you have to appeal to their emotions—connect with something they care about. Their worries, their fears. Or show them a compelling vision of the future: give a human example. Walk through how a real person will experience this product—their day, their family, their work, the change they'll experience. Just don't lean so far into the emotional connection that what you're arguing for feels novel, but not necessary.

There's an art to telling a compelling story. But there's also a science.

And always remember that your customers' brains don't always work like yours. Sometimes your rational argument will make an emotional connection. Sometimes your emotional story will give people the rational ammunition to buy your product. Certain Nest customers looked at the beautiful thermostat that we lovingly crafted to appeal to their heart and soul and said, "Sure, okay. It's pretty" and then had a thrilled, emotional reaction to the potential of saving twenty-three dollars on their energy bill.

Every person is different. And everyone will read your story differently.

That's why analogies can be such a useful tool in storytelling. They create a shorthand for complicated concepts—a bridge directly to a common experience.

That's another thing I learned from Steve Jobs. He'd always say that analogies give customers superpowers. A great analogy allows a customer to instantly grasp a difficult feature and then describe that feature to others. That's why "1,000 songs in your pocket" was so powerful. Everyone had CDs and tapes in bulky players that only let you listen to 10–15 songs, one album at a time. So "1,000 songs in your pocket" was an incredible contrast—it let people visualize this intangible thing—all the music they loved all together in one place, easy to find, easy to hold—and gave them a way to tell their friends and family why this new iPod thing was so cool.

Everything at Nest was steeped in analogies. They filled our website and our videos and our ads and even our support articles, our installation guides. They had to. Because to truly understand many of the features of our products, you'd need a deep well of knowledge about HVAC systems and power grids and the way smoke refracts through a laser to detect fire—knowledge almost nobody had. So we cheated. We didn't try to explain everything. We just used an analogy.

I remember there was one complex feature that was designed to lighten the load on power plants on the hottest or coldest days of the year when everyone cranked up the heat or AC at once. It usually came down to just a few hours in the afternoon, a few days a year—one or more coal power plants would be brought on line to avoid blackouts. So we designed a feature that predicted when these moments would come, then the Nest Thermostat would crank the AC or heat up extra before the crucial peak hours and turn it down when everyone else was turning it up. Anyone who signed up for the program got a credit on their energy bill. As more and more people joined the program, the result was a win-win—people stayed comfortable, they saved money, and the energy companies didn't have to turn on their dirtiest plants.

And that is all well and good, but it just took me 150 words to explain. So after countless hours of thinking about it and trying all the possible solutions, we settled on doing it in three: Rush Hour Rewards.

Everyone understands the concept of a rush hour—the moment when way too many people get on the road together and traffic slows to a creep. Same thing happens with energy. We didn't need to explain much more than that—rush hours are a problem, but when there's an energy rush hour, you can get something out of it. You can get a reward. You can actually save money rather than getting stuck with everyone else.

We made a whole webpage about it with a car graphic and little power plants puffing away. We probably belabored the point and stretched the metaphor—but we knew most people wouldn't dig in that far.

For the vast majority of customers, we made it simple. With three words and one analogy we helped people get it—when there's an energy rush hour, your Nest Thermostat can save you money.

It's a story. A very quick one, but that's the best kind.

Quick stories are easy to remember. And, more importantly, easy to repeat. Someone else telling your story will always reach more people and do more to convince them to buy your product than any amount of talking you do about yourself on your own platforms. You should always be striving to tell a story so good that it stops being yours—so your customer learns it, loves it, internalizes it, owns it. And tells it to everyone they know.

Chapter

3.3

EVOLUTION VERSUS DISRUPTION VERSUS EXECUTION

Evolution: A small, incremental step to make something better.

Disruption: A fork on the evolutionary tree—something fundamentally new that changes the status quo, usually by taking a novel or revolutionary approach to an old problem.

Execution: Actually doing what you've promised to do and doing it well.

Your version one (V1) product should be disruptive, not evolutionary. But disruption alone will not guarantee success—you can't ignore the fundamentals of execution because you think all you need is a brilliant disruption. And even if you do execute your idea well, it may not be enough. If you're revolutionizing a major, dug-in industry, you may also need to disrupt marketing or channel or manufacturing or logistics or the business model or something else that never occurred to you.

Assuming V1 was at least a critical success, the second version of your product is typically an evolution of your first. Refine what you made in V1 using data and insights from actual customers and double down on your original disruption. The execution should step up a notch—now you know what you're doing and should be able to provide a significantly more functional product.

You can continue evolving that product for a while, but always seek out new ways to disrupt yourself. You can't only start thinking about it when the competition threatens to catch up or your business begins to stagnate.

••

If you're going to pour your heart into creating something new, then that thing should be disruptive. It should be bold. It should change *something.* It doesn't have to be a product—Amazon was a disruptive service long before they got into making their own hardware. You can disrupt how things are sold, delivered, serviced, financed. You can disrupt how they're marketed or recycled.

Disruption should be important for you personally—who doesn't want to do something exciting and meaningful?—but it's also important for the health of your business. If you've truly made something disruptive, your competition probably won't be able to replicate it quickly.

The key is to find the right balance—not so disruptive that you won't be able to execute, not so easy to execute that nobody will care. You have to choose your battles.

Just make sure you have battles.

If you undershoot—if you create something that's only evolutionary, just another step down a well-trod path—then when you pitch it to the smartest people you know across a range of disciplines, they'll just shrug. "Eh. Okay."

There'll be an almost audible fizzle.

You need something that will make them stop in their tracks and say, "Wow. Tell me more." Whatever you're disrupting is going to be the thing that defines your product—the thing that will make people take notice.

And it will be the thing that will make them laugh. If you're disrupting big, entrenched industries, your competition will almost certainly dismiss you in the beginning. They'll say that what you're making is a plaything, not a threat. They'll flat out laugh in your face.

Sony laughed at the iPod. Nokia laughed at the iPhone. Honeywell laughed at the Nest Learning Thermostat.

At first.

In the stages of grief, this is what we call Denial.

But soon, as your disruptive product, process, or business model begins to gain steam with customers, your competitors will start to get worried. They'll start paying attention. And when they realize you might steal their market share, they'll get pissed. Really pissed. When people hit the Anger stage of grief, they lash out, punch something. When companies get angry they undercut your pricing, try to embarrass you with advertising, use negative press to undermine you, put in new agreements with sales channels to lock you out of the market.

And they might sue you. If they can't innovate, they litigate.

The good news is that a lawsuit means you've officially arrived. We had a party the day Honeywell sued Nest. We were thrilled. That ridiculous lawsuit (they sued our thermostat for being round) meant we were a real threat and they knew it. So we brought out the champagne. That's right, fuckers. We're coming for your lunch.

We had no intention of folding. We knew that for decades Honeywell had been suing little innovative companies out of business. They'd put a noose around their necks until the upstarts had no choice but to sell to Honeywell for a pittance. Any threat was quickly stamped out. But Nest's general counsel, Chip Lutton, and I had been through these wars together from the Apple days and were not about to get spooked into settling. [See also: Chapter 5.7: Lawyer Up: The first time I had to deal with a lawsuit.]

If your company is disruptive, you have to be prepared for strong reactions and stronger emotions. Some people will absolutely love what you've made. Some people will violently, relentlessly hate it. That's the danger with disruption. It is not welcome by everyone. Disruption makes enemies.

Even starting something new in a big company won't protect you. You'll have to deal with politics, jealousy, and fear. You're trying to change things, and change is scary, especially to people who think they've mastered their domain and who are completely unprepared for the ground to shift under their feet.

All it takes to start a landslide is one big scary new thing. Maybe two.

Just don't overshoot. Don't try to disrupt everything at once. Don't make the Amazon Fire Phone.

I remember when Jeff Bezos first mentioned the idea. We were having a breakfast meeting about me potentially joining Amazon's board. Jeff hinted at plans for making a new line of Amazon-branded devices, in particular a phone. It would be spectacularly disruptive: everything would look 3-D, it would let you X-ray any media, you could scan anything in the world and then buy it on Amazon. It would change everything.

I told him he'd already disrupted hardware—with the Kindle. It was wonderfully innovative and had a unique platform that nobody could duplicate. To get Amazon onto people's phones and change how people shopped online, you didn't need to build a whole new device. You just needed a really great app that would live on everybody else's devices.

I told him: I wouldn't make the phone.

He made the phone.

I didn't get the board seat.

When it launched, the Fire Phone did everything he had promised—but none of it well. They tried to do too much, change too much. So the disruptions turned into gimmicks, and the project failed. It was a hard, painful lesson that they haven't made again. Do. Fail. Learn.

But that's the tricky thing with disruptions—they're an extremely delicate balancing act. When they fall apart it's usually for one of three reasons:

1. You focus on making one amazing thing but forget that it has to be part of a single, fluid experience. [See also: Figure 3.1.1, in Chapter 3.1.] So you ignore the million little details that aren't as exciting to build—especially for V1—and end up with a neat little demo that doesn't actually fit into anyone's life.
2. Conversely, you start with a disruptive vision but set it aside because the technology is too difficult or too costly or doesn't work well enough. So you execute beautifully on everything else but the one thing that would have differentiated your product withers away.
3. Or you change too many things too fast and regular people can't recognize or understand what you've made. That's one of the (many) issues that befell Google Glass. The look, the technology—it was all so new that people had no idea what to do with it. There was no intuitive understanding of what the thing was for. It's as if Tesla decided out the gate to build electric cars with five wheels and two steering wheels. You can change the motor, change the dash—but it still has to look like a car. You can't push people too far outside their mental model. Not at first.

That third reason explains why the first-generation iPod didn't have the iTunes music store. There was no music marketplace, the term "podcast" was months away, and users just ripped their CDs with iTunes or pirated them online.

And it's not because we didn't think of it. We dreamed about various iTunes features while we were building the iPod. But we didn't have time to execute it and we'd already disrupted enough. We needed to get people from CDs to MP3s—that was a big leap already. We'd only be successful if they had time to catch their balance before we asked them to jump again.

As we started working on V2 and V3, adding a digital marketplace became the logical next step. We were maximizing and capitalizing on our initial disruption. There was plenty of low-hanging fruit, so we just kept refining, evolving. V4, V5, V6.

The more we evolved, the more we wanted to change. At one point we came to Steve with radically new designs that we were really excited about—they were smaller and lighter, innovative and beautiful. And we'd removed the click wheel. He looked at them and said, "They're great. But you've lost what it means to be an iPod."

The world saw the click wheel and thought "iPod." So removing it wasn't an evolution—it was a disruption that made no sense at that moment. If we'd followed through we'd have made a smaller, lighter music player and diminished our own brand.

Lesson learned.

When you're evolving you need to understand the quintessential things that define your product. What's key to your feature set and your branding? What have you trained the customer to look for? With the iPod it was the click wheel. With the Nest Learning Thermostat it was the round, clean screen with a big temperature in the middle.

To maintain the core of your product there are usually one or two things that have to stay still while everything else spins and changes around them.

And that's a useful constraint. You need some constraints to force you to dig deep and get creative, to push envelopes you hadn't thought to open before.

At Apple we pushed ourselves nonstop. We knew that we needed to launch a new, significantly improved iPod every single year, ready to be gifted for the holidays. This was the first time Apple had set that kind of pace, since Mac products were always driven around computer processor upgrades from our suppliers. [See also: Chapter 3.5: Heartbeats and Handcuffs: For a long time Macintoshes were at the whim of IBM.] But in our heads we heard the sound of Sony and our other competitors stalking us. We were in the lead but we had to keep evolving and executing pristinely to stay that way. Each year's iPod had to be substantively better than last year's model—either the hardware or software or both. We needed to keep the competition at bay and give customers a reason to upgrade.

So we learned to underpromise and overdeliver. We'd be conservative about key features like battery life—all through development we'd make sure we'd reached a number that Steve was satisfied with. Thirteen hours, fourteen hours. But behind the scenes, we'd steadily try to improve it—to claw back a minute here, a minute there.

Then we'd launch the latest iPod with the latest specs: fourteen hours of battery life.

Reviewers would get their hands on the new iPod and not only would it deliver, it would overdeliver. It would run hours longer than they expected.

We did it over and over, year after year, but somehow nobody caught on. Every time was a surprise. A delight. And that did just as much to cement Apple's reputation for excellence as the iPod's design and user experience.

That relentless march did a lot to define the iPod brand and keep people's attention on Apple. And it depressed the hell out of our competition. I had friends at Philips who told me that every time they'd think of a great idea for how to outmaneuver the iPod, we'd come out with a similar feature a few months later and they'd have to go back to the drawing board. It crushed their spirits. We were moving so fast that by the time they caught up they were already behind.

But you can only evolve for so long.

Eventually the competition started closing in. The iPod had blown every other MP3 player out of the water, we had over 85 percent worldwide market share—but ultracompetitive cell phone manufacturers began figuring out how to get a piece of our pie. They started putting MP3s onto their phones, seeing the potential of getting everything into one device: calls, texts, the snake game, *and* music.

At the same time, the world was adopting cell phones like crazy and data networks were becoming dramatically better, faster, and cheaper. It was obvious that soon enough most people would be able to stream music instead of downloading it. That would change everything for the iPod business.

So either the landscape was going to change under our feet, or we were going to change the landscape.

We had to disrupt ourselves.

The iPod was Apple's only successful new non-Mac product in fifteen years. At times it accounted for more than 50 percent of Apple's revenue. It was hugely popular and still growing fast. It defined the company for millions of non-Mac customers.

But we decided to eat our own. We had to make the iPhone, even though we knew it could, probably would, kill the iPod.

It was an enormous risk. But with any disruption, the competition only wallows in Denial and Anger for so long. Eventually they reach Acceptance, and if they've got any life left in them, they'll start working furiously to catch up to you. Or you may inspire a whole new wave of companies who can use your initial disruption as a stepping-stone to leapfrog you.

When you can see the competition nipping at your heels, you have to do something new. You have to fundamentally change who you are as a business. You have to keep moving.

You cannot be afraid to disrupt the thing that made you successful in the first place. Even if it made you hugely successful. Look at Kodak. Look at Nokia. Companies that become too big, too comfortable, too obsessed with preserving and protecting that first big innovation that put them on the map—they topple. They unravel. They die.

If you're experiencing your biggest market share ever, that means you're on the brink of becoming calcified and stagnant. It's time to dig deep and kick your own ass. Google, Facebook, all the tech giants are due for a disruption any day now—or they'll be forced into it by regulation.

Tesla could have fallen into the same trap. They started with one huge disruption—revolutionizing the auto industry, making EV cars attractive to consumers for the first time. But as every carmaker in the world followed their lead, Tesla was in danger of becoming just one more electric car in a market flooded by them. So they started electrifying different kinds of vehicles and innovating charging networks and retail and service, batteries and supply chains. They're ensuring that the competition has to fully disrupt every part of their operations to even enter the race. Once every carmaker has an electric vehicle, then the customer will focus on all these other aspects that Tesla has already disrupted and brought to market.

Competition is a given, both direct and indirect. Someone is always watching, trying to exploit any crack in a more successful competitor.

For years Microsoft's primary source of revenue was selling Windows to giant corporations. It was a sales-driven culture, not product-driven. So the product didn't change much, year after year, long after the internet was born and started to change everything else. Long after it was clear Microsoft's business model was dying. Long after the company culture sank into a deep malaise and the industry dismissed them as a dinosaur.

But eventually, after years of flailing, the new CEO, Satya Nadella, shook up their culture and forced them to look at other products and business models. They branched out. And they had plenty of false starts—plenty of failed products. Many branches broke—but several bore fruit: the Surface products, Azure cloud computing. They stopped looking to Windows to be the cash cow and turned Office into an online subscription. They climbed out of their hole, their stagnant swamp, and now Microsoft is back to making innovative imagination-grabbing products again, like Hololens and their Surface products.

Of course, most founders would kill to start a business that gets so large that it's at risk of stagnation. Almost nobody gets that far.

Most people stall out at the first step—the first disruption. It's easy to say "change something meaningful," and infinitely harder to come up with a great idea and execute it in a way that connects with customers. [See also: Chapter 4.1: How to Spot a Great Idea.]

Especially because one amazing disruption may not be enough. You may have to disrupt things you never even thought of.

If Nest had only disrupted hardware—if we had just built the Nest Learning Thermostat alone—we would have failed. Utterly.

We needed to disrupt the sales and distribution channel, too.

At the time, regular people didn't really go out and buy thermostats. You could get them at a hardware store, but they were intentionally complicated so you couldn't install them easily yourself. And they weren't sold online so you couldn't comparison shop and see the steep markup HVAC technicians would charge you. So if your thermostat broke, you'd just call a technician to replace it. And if your heater or AC conked out you'd be upsold a new thermostat, too (whether you needed it or not).

For every upsell of a fancy new Honeywell thermostat, the HVAC technician would get a nice little bonus for a job well done. Sell enough thermostats and Honeywell would send you on a Hawaiian vacation.

This was an entrenched market where the existing players had done everything in their power to keep out competitors. There was no incentive for HVAC technicians to sell or install Nest Learning Thermostats. We weren't giving out bonuses—in fact they'd make less money selling our devices than the old ones. And we definitely weren't going to send anyone to Hawaii. We were a tiny company and Honeywell had bought the installers' loyalty for decades.

So we had to go completely around the existing channel. We had to create a new market: selling directly to homeowners in a world where homeowners did not buy thermostats. And we had to sell in places where thermostats had never been sold before. Our first retail partner was Best Buy, and they had no idea where to put Nest. It's not like they had a thermostat aisle.

But I made sure we were not going to repeat the mistakes of Philips. We weren't going to let Nest get shoved behind a stereo in a storeroom somewhere. So we told Best Buy that we didn't want a thermostat aisle—we wanted a Connected Home aisle. Of course they didn't have that, either. So we invented it together.

I didn't get into the thermostat business to disrupt Best Buy. But that's what we had to do to sell thermostats.

If you do it right, one disruption will fuel the next. One revolution will domino into another. People will laugh at you and tell you it's ludicrous, but that just means they're starting to pay attention. You've found something worth doing. Keep doing it.

Chapter

3.4

YOUR FIRST ADVENTURE— AND YOUR SECOND

When you're leading a team or project to launch V1—the first version of a product that's new for you and your team—it's like heading out into the mountains with friends for the first time. You think you have everything you need to camp and climb, but you've never done it before. So you're tentative. And you're slow. But you take your best guess at what you need and where you're going and head into the wild.

The next year, you decide to do it again. This time it's V2. And it's completely different—you know where you're going, you know what it takes to get there, and you know your team. You now have the confidence to be bolder, take bigger risks, to go further than you ever thought before.

But on that very first journey, you won't have those advantages. You'll need to make many opinion-driven decisions without the benefit of data or experience to guide you. [See also: Chapter 2.2: Data Versus Opinion.]

The tools you need to make those decisions are below, organized by order of importance:

1. **Vision:** Know what you want to make, why you're making it, who it's for, and why people will buy it. You'll need a strong leader or a small group to ensure the vision is delivered intact.

2. **Customer insights:** This is what you've learned through customer or market research or simply by thinking like your customer: what they like, what they dislike, what problems they experience on a regular basis, and what solutions they'll respond to.

3. **Data:** For any really new product, reliable data will be limited or nonexistent. That doesn't mean you shouldn't make a reasonable attempt to gather objective information—the scope of the opportunity, the way people use current solutions, etc. But this information will never be definitive. It won't make your decisions for you.

Once you start iterating on an existing product, V2, your second adventure, you'll have experience and customers and the luxury of plenty of data-driven decisions. However, a myopic focus on numbers can slow you down or lead you off-track. So you'll still need all the same tools as above, just in a different order.

1. **Data:** You'll be able to track how customers use your current product and test new versions. You can confirm or disprove hunches with hard data from actual paying customers. This data will allow you to fix the stuff you screwed up when you were just following your gut.

2. **Customer insights:** Once people have committed to paying money for your product, they're much more reliable for useful insights. They can tell you what's broken and what they want to see next.

3. **Vision:** Assuming you got 1.0 more or less right, that original vision moves behind the data and insights you can get from actual customers. But your original vision should not be set aside entirely as you iterate. You should always keep in mind your longer-term

goals and mission so that your product's fundamental purpose doesn't get lost.

You should also keep in mind that you're not just making V1 or V2 of your product—you're building out the first or second version of your team and processes.

V1 team: It's mostly or all new players working together. You're still feeling each other out, trying to understand if you can trust each other and who will stick around when things get hard. You'll need to agree on a singular process, which is often harder than agreeing on a product. People will disagree based on past experiences and trust can quickly break down. The risk of making something new is always compounded by not having confidence in the team.

V2 team: You may have to upgrade parts of your team as you become more ambitious, but many of the same teammates who weathered the storm of V1 will be ready to enter the fray again for V2. You'll hopefully trust each other, have already settled into a development process that works, and you'll have a shorthand that speeds everything up. This confidence in each other will allow you to take bigger risks and build more exciting products.

••

The marketing team fought Steve Jobs the hardest about the iPhone keyboard. But a lot of us rebelled. In 2005 the most popular "smart" phone by far was the BlackBerry—fondly known as the Crackberry. People were addicted. BlackBerry owned 25 percent of the market and was growing fast. And the BlackBerry die-hards would always tell you that the very best thing about their very favorite gadget was obvious. It was the keyboard.

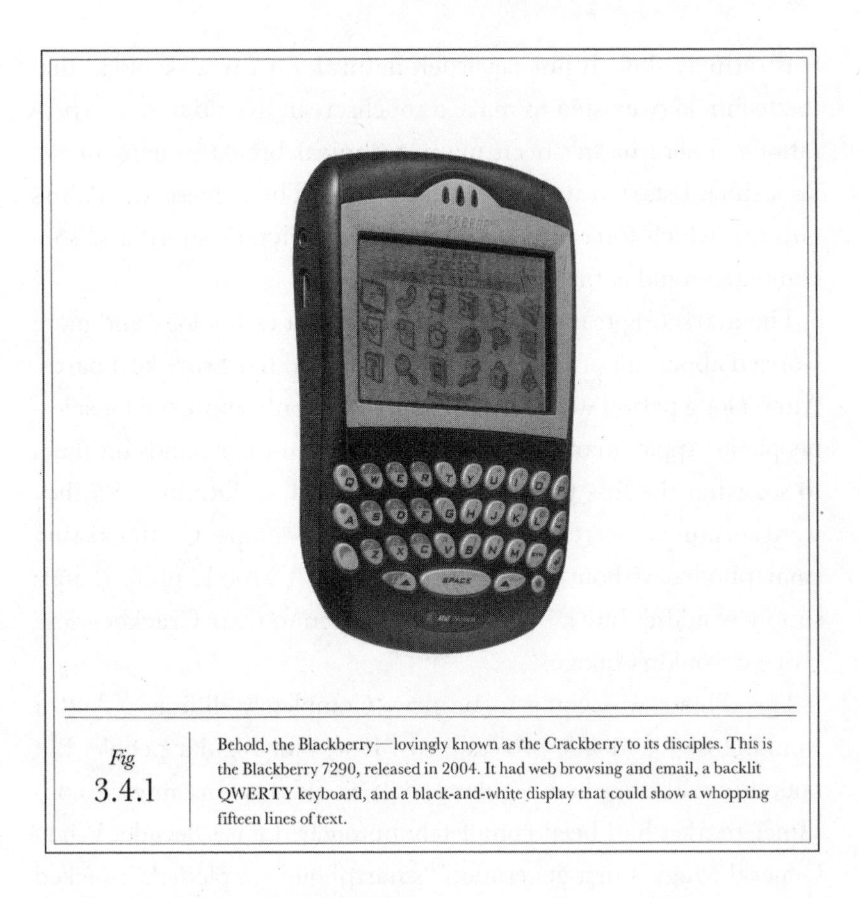

Fig 3.4.1 Behold, the Blackberry—lovingly known as the Crackberry to its disciples. This is the Blackberry 7290, released in 2004. It had web browsing and email, a backlit QWERTY keyboard, and a black-and-white display that could show a whopping fifteen lines of text.

It was built like a tank. It took a couple of weeks to get used to, but after that you could text and email incredibly fast. It felt good under your thumbs. Solid.

So when Steve told the team his vision for Apple's first phone—one giant touchscreen, no hardware keyboard—there was an almost audible gasp. People whispered in the hallways, "Are we really going to make a keyboardless phone?"

Touchscreen keyboards sucked. Everyone knew they sucked. I *really* knew they sucked. I'd built them twice—first at General Magic and then at Philips. You'd have to use a stylus, tap-tap-tapping away at a screen with no give, no feedback, slipping and sliding, frustratingly slow. It just never felt natural. So I was skeptical that the technology existed to make a touchscreen live up to our expectations. There hadn't been many technical breakthroughs in the area since I started working on it in 1991. The biggest was Palm's Graffiti, which forced you to write in hieroglyphic shorthand so a computer could actually understand it.

The marketing team was less worried about technology and more worried about sales. They *knew* people wanted hardware keyboards. After a long period where BlackBerrys were only approved for salespeople at Apple, marketing had finally gotten their hands on them to see what the fuss was about. And they fell in love, too. So they were certain we weren't going to be able to compete with existing smartphones without a hardware keyboard. Mobile professionals simply wouldn't buy it—they were addicted to their Crackberries.

Steve wouldn't budge.

The iPhone was going to be new. Completely different. And it wouldn't be for mobile professionals. It was for regular people. But nobody could know how regular people would react because the consumer market had been completely untouched for a decade. When General Magic's first-generation "smartphone" imploded, it sucked up the entire industry's will to build personal devices for Joe Sixpack.

Most of the hardware manufacturers in the 1990s and early 2000s did what I did: they turned to business tools. Philips, Palm, BlackBerry. They all targeted businesspeople who mostly needed to write emails, send messages, and update docs. Not watch movies. Not listen to music. Not screw around on the internet or take pictures or connect with friends.

And the iPhone was going to be tiny—Apple didn't want it much bigger than an iPod, so it could easily slip in and out of your pocket. Ultimately the screen measured 3.5 inches diagonally. And Steve wasn't about to sacrifice half that space to a molded plastic keyboard that was impossible to change without literally going back to the drawing board.

Fig 3.4.2 The original iPhone launched in 2007 was tiny—smaller than any iPhone you can get today. It was 4.53 x 2.40 inches, weighed 135 grams, and had a 3.5-inch screen. By comparison, the iPhone 13 mini measures 5.8 x 2.53 inches, weighs 141 grams, and boasts a 5.4-inch screen.

A hardware keyboard locks you into a hardware world. What if you want to write in French? Or Japanese? Or Arabic? What if you want emojis? What if you need to add a function or take one away? And what if you want to watch a video? There's no turning a phone horizontally if it's half keyboard.

Fig 3.4.3

It's easy to see Steve's point when you compare the Blackberry Curve 8310 (released in August 2007) to the original iPhone (released in June 2007). The Blackberry's screen was only 2.5 inches. Its keyboard was so robust that almost no screen remained.

I agreed with Steve. In principle. I just didn't think we could pull it off with any of the technology I'd seen to date. I needed enough data to know that I could make his vision a reality. So to get comfortable and stop arguing opinions, we set down weekly challenges to the hardware and software teams to create a better demo. How fast could we get it? What was the error rate? The keys were going to be smaller than your fingers, so errors were inevitable. How were we going to get around those errors and correct them? And at what speed? When was each key activated—when you put your finger down or when you lifted it up? How was it going to sound? We needed audio feedback if you couldn't get force feedback. And then there was the qualitative test: Did it feel good? Did I want to use it? Did it drive me crazy? We had to change the algorithms at all levels of the system many, many, many times.

After eight weeks it was far from perfect, but it was getting there. Considering how much we'd improved in just a couple of months, I decided that, though it wouldn't be as good as a hardware keyboard, it would be just good enough. I convinced myself.

But marketing wasn't moved.

After weeks and weeks of argument, Steve put his foot down. There was no data that would definitively prove it would work. There was no data that would prove it wouldn't. This was an opinion-driven decision and Steve's opinion was the one that counted most. "So either get on board right now or you're off the team," Steve said. That settled it for marketing.

Of course, in the end Steve was proven right. The iPhone changed everything. And it was only possible because he stuck to his vision.

But that's not to say that sticking to your vision will always lead to success.

Not even for Steve Jobs.

Most people don't realize what the iPod was originally built for. Its purpose wasn't just to play music—it was made to sell Macintosh computers. That's what was in Steve Jobs's head: We're going to make something amazing that will only work with our Macs. People will love it so much that they'll start buying Macs again.

At the time Apple was near death. It had almost no market share—even in the United States. But the iPod would solve that problem. It would save the company.

So as far as Steve Jobs was concerned, the iPod would never work with a PC. That would completely defeat the point—we needed to sell more computers

And that's why the first generation of iPod fizzled.

The critics loved it. So did the people who already had Apple computers. Unfortunately, at that time there weren't many of them. The iPod cost $399. The entry-level iMac cost $1,300. Even though the iPod was by far the best MP3 player on the market, nobody was going to drop $1,700 on the complete Apple package just to listen to Radiohead more easily.

But that didn't stop us. The same day we introduced the first version, we had already started working on the second. V2 would be thinner, more powerful, more beautiful. We came to Steve and said it needed to work on PCs. It had to.

No.

Absolutely not.

Forcing Steve to abandon his original plan was almost impossible. But we waged an all-out war trying to prove to him that this wasn't an opinion-driven decision anymore. This was data. We were in V2. We had real revenue and insights from actual paying customers (although not enough of either).

We were iterating. We were climbing the mountain again. It was time to put the vision third.

We managed to get him to consider a half measure for the second-generation iPod—adding the Musicmatch Jukebox (basically the leading iTunes competitor but on a PC), which let you transfer your music library from a Windows machine to the iPod. And even that was a struggle.

In the end, we agreed that we should ask Walt Mossberg, the famed tech reviewer, to cast the deciding vote (unbeknownst to Walt!). It was a setup, of course. I think Steve wanted someone to blame if it didn't work.

Ultimately Steve was proven wrong. Allowing the iPod to work with PCs instantly buoyed sales. By the third generation we started selling tens of millions. Then hundreds of millions. That's what turned things around for Apple. That's what saved the company. Ironically, it's even what saved the Mac—customers who loved the iPod started looking into Apple's other products and Macintoshes started selling again.

But the lesson here isn't that Steve Jobs was fallible. Of course he was. He was human.

The lesson is about when and how vision and data should guide your decisions. In the very beginning, before there are customers, vision is more important than pretty much anything else.

But you don't have to figure out your vision all by yourself. In fact, you probably shouldn't. Locking yourself alone in a room to create a manifesto of your single, luminous vision looks and feels indistinguishable from completely losing your mind. Get at least one person—but preferably a small group—to bounce ideas off of. Sketch out your mission together. Then fulfill it together.

In the end, you may create something magical and world-changing. But then again, you might not.

There's always a chance that you valiantly clung to your vision for 1.0 in the face of all obstacles and the vision turned out to be wrong. [See also: Chapter 3.6: Three Generations: The "chasm" is the hole companies can fall into.] Whatever you made just doesn't work. Maybe there was a data-driven decision that you thought was an opinion. Maybe you just calculated wrong or timed it wrong or something changed in the macro environment that you couldn't control.

At that point you have to go back and, as painful as it is, honestly and thoroughly analyze why you failed. This is the moment when you need to gather data. Your gut got you to this point, so find data to help you understand why your gut was wrong.

You may not come back from it. You may have run out of money, lost the team or your credibility. But the only way to move forward is to do an honest accounting of the past. Learn your lessons—especially the hard ones. Then try again. Back to the drawing board. V1.

Eventually your vision will improve. You'll learn to trust your gut again. And you'll get to the other side: V2. And that's a very different story.

When you're building the second version of your product, you can talk to actual customers and understand exactly what they think and what they want to see next. You can do all the stuff you desperately wanted to put into V1 but couldn't. You can analyze the numbers, understand the costs and benefits. You can confirm your insights with information, A/B tests, charts, and figures. You can adjust and adapt to your customer's needs and more and more decisions can be driven by glorious, simple, clear-cut, black-and-white data.

But before that moment comes you need to get through the sprint and the marathon of V1. You need people you trust to keep you going. And you'll need to know when to stop.

If you wait for your product to be perfect, you will never finish. But it's very hard to know when you're done—when you need to stop building and just put it out into the world. When is it good enough? When are you close enough to your vision? When are the inevitable issues ignorable enough that you can live with them?

Typically your vision is so much greater than what materializes in V1. There's always another revision, always something else you want to do, change, add, tweak. When do you tear yourself away from what you're making and just . . . stop? Ship it. Set it free. See what happens.

Here's the trick: write a press release.

But don't write it when you're done. Write it when you start.

I began doing this at Apple and eventually realized other leaders had figured it out, too (looking at you, Bezos). It's an incredibly useful tool to narrow down what really matters.

To write a good press release you have to focus. The press release is meant to hook people—it's how you get journalists interested in what you're making. You have to catch their attention. You have to be succinct and interesting, highlight the most important and essential things that your product can do. You can't just list everything you want to make—you have to prioritize. When you write a press release you say, "Here. This. This is what's newsworthy. This is what really matters."

So spend some time developing as great a press release as you can. Consult with marketing and PR people if you have to. They'll help you trim it down to the essentials.

Then weeks or months or years later, as you're getting close to finishing, as you're debating what makes it in, what gets cut, what matters, what doesn't—take out your press release. Read it.

If you launched right now, could you more or less send that press release into the world and have it be mostly true? If the answer is yes, then congratulations: your product is probably ready, or at least pretty close. You have achieved the core of your vision. Everything else is most likely a "nice to have," not a priority.

Of course, there's a chance that since you started, you've had to pivot so much that the original press release is laughably off-base. Sometimes that happens.

No problem. Write another press release. Rinse. Repeat.

This is an adventure and adventures never go according to plan. That's what makes them fun. And scary. And worth doing. That's why you take a deep breath, surround yourself with great people, and head out into the wilderness.

Chapter 3.5

HEARTBEATS AND HANDCUFFS

You need constraints to make good decisions and the best constraint in the world is time. When you're handcuffed to a hard deadline, you can't keep trying this and that, changing your mind, putting the finishing touches on something that will never be finished.

When you handcuff yourself to a deadline—ideally an external, immovable date like Christmas or a big conference—you have to execute and get creative to finish on time. The *external heartbeat*, the constraint, drives the creativity, which fuels the innovation.

Before you launch V1, your external deadline is always a little wobbly. There are too many unknown unknowns to write it in stone. So the way you keep everyone moving is by creating strong internal deadlines—heartbeats that your team sets their calendar to:

1. **Team heartbeats:** Each individual team makes its own rhythm and deadlines for delivering their piece of the puzzle. Then all the teams align for . . .

2. **Project heartbeats:** These are the moments when different teams sync to make sure the product still makes sense and all the pieces are moving at the right pace.

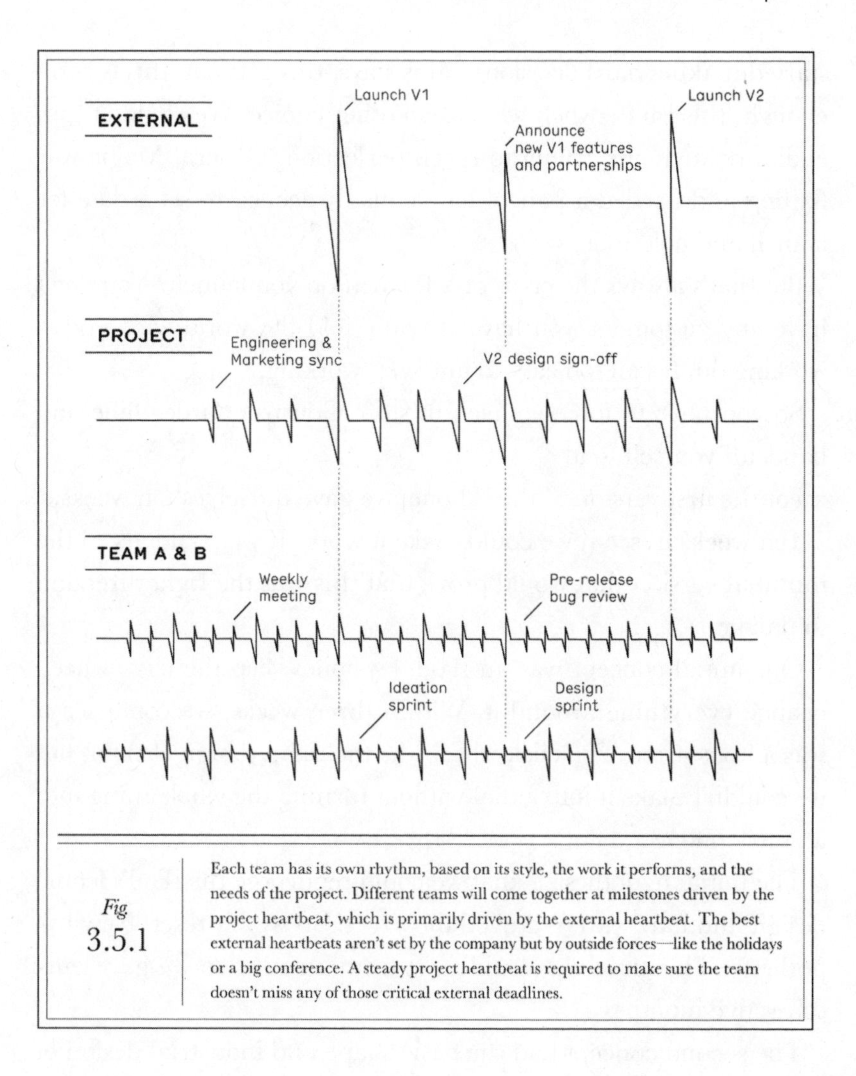

Fig 3.5.1 Each team has its own rhythm, based on its style, the work it performs, and the needs of the project. Different teams will come together at milestones driven by the project heartbeat, which is primarily driven by the external heartbeat. The best external heartbeats aren't set by the company but by outside forces—like the holidays or a big conference. A steady project heartbeat is required to make sure the team doesn't miss any of those critical external deadlines.

••

When I joined General Magic, the plan was to ship in nine months. Then the launch got delayed by six months. And another six months. And another. It kept going like that for four years.

Possibly the only reason we shipped at all was that Apple launched the Newton and investors started putting the pressure on. That's when we encountered our very first constraint: the competition closing in.

The Magic Link only launched when it had to launch. We only

started making hard decisions—this stays, this gets cut, this is good enough, this isn't—when we had no other choice. We couldn't spin endlessly anymore, trying to reach perfection. General Magic was flailing and needed a pair of handcuffs. It needed to set a date for launch and hold to it.

But that's always the crisis of V1: when do you launch? You don't have any customers, you haven't really told the world what you're working on. It's all too easy to just keep working.

So you have to force yourself to stop. Construct a deadline and handcuff yourself to it.

For the first version of the iPhone, we gave ourselves ten weeks.

Ten weeks to see if we could make it work. If we could get to the minimal version that would prove that this was the right direction to pursue.

Our initial concept was an iPod + phone: keep the click wheel, change everything around it. Within three weeks, we could see it was a nonstarter. The click wheel was the main design element but we couldn't make it into a dial without turning the whole thing into a rotary phone.

The initial hypothesis—that we could repurpose the iPod's iconic design and hardware—proved incorrect. So we hit reset. New hypothesis. This time we would start from scratch, so we gave ourselves five months.

The second concept had the basic shape and industrial design of an iPod Mini but with a full face screen and no click wheel—very similar to what we see today.

We ran into a slew of new problems with that second iPhone prototype design—we couldn't get the engineering right. The antennas, the GPS, the cameras, the thermals. We'd never built a mobile phone before, let alone a smartphone, and our assumptions were flawed. Again.

Reset. Start over.

Fig. 3.5.2

This iPod-phone model wasn't actually one of ours—it arrived from a manufacturer who'd heard rumors that we were working on a phone and wanted to pitch us their idea. This odd-looking device shows the impossibility of designing a phone around the click wheel. The top half would swivel around 180 degrees so you could use the screen if you were dialing a phone number or texting—not a bad idea, actually, but it's not an iPhone.

It was only with the third version that we understood all the pieces well enough to create the right V1 device.

But we would have never reached that third design if we hadn't given ourselves hard deadlines with the first two—if we hadn't cut ourselves off after a few months, reset, and moved on.

We forced as many constraints on ourselves as possible: not too much time, not too much money, and not too many people on the team.

That last point is important.

Fig 3.5.3 We spent a lot of time playing with various concepts, and these crazy plastic slabs were from some early form factor trials. They let us see how our ideas would feel in our hands, pockets, and purses as we tried to figure out what made sense and what didn't.

Don't go crazy hiring people just because you can. With most projects in the concepting stage, you can get a huge amount done with around ten people or even fewer. You don't want to staff up and then be forced to design by committee or put a ton of people on the sidelines, sitting on their hands, waiting for you to figure it out.

By the end of the first iPhone project we had about eight hundred people working on it. But can you imagine what would have happened if all eight hundred had been with us from the beginning, watching us abandon the vision and restart the project? And then do it again a few months later? It would have been mayhem. Eight hundred people panicking and us endlessly reassuring everyone, focusing on the positive, trying to keep all those people in sync with a truly crazy number of iterations.

So keep your project small as long as you can. And don't allocate too much money at the start. People do stupid things when they have a giant budget—they overdesign, they overthink. That inevitably leads to longer runways, longer schedules, and slower heartbeats. Much, much slower.

Generally any brand-new product should never take longer than 18 months to ship—24 at the limit. The sweet spot is somewhere between 9 and 18 months. That applies to hardware and software, atoms and bits. Of course, there are things that take longer—research can take decades, for example. But even if it takes ten years to research a question, regular check-ins along the way ensure you're still chasing the right answer. Or still asking the right question.

Every project needs a heartbeat.

Pre-V1 launch, that heartbeat is entirely internal. You're not talking to the outside world yet, so you have to have a strong internal rhythm that pushes you toward a set launch date.

This rhythm is made of major milestones—board meetings, all-hands meetings, or project milestones at certain moments of product development where everyone, engineering and marketing and sales and support, can pause and sync with each other. This might happen every few weeks or every few months, but it has to happen in order to keep everyone moving in lockstep to the external announcement.

And in order to keep the project heartbeat going, every team will need to produce its own deliverables at its own pace. Each team's heartbeat will be different—it could be six-week sprints or weekly reviews or daily check-ins. It could be scrum or waterfall or kanban, whatever organizational framework or project management approach works for you. A creative team is going to have a very different heartbeat than an engineering team; a company that makes hardware is going to have slower team rhythms than companies that only push around electrons. It doesn't matter what that heartbeat is, your job is to keep it steady so your team knows what's expected of them.

I learned that at Philips, the first time I had to create a heartbeat from scratch.

When we started, the whole team was pretty young and inexperienced at project management, so we hired some consultants to help us build a schedule. They suggested we organize our tasks down to half days. The team would estimate how many half days it would take to complete each piece of the project and we'd break down all the months, weeks, and days necessary to do every task we could think of. Then we created detailed schedules 12–18 months into the future based on everyone's individual workload.

It seemed perfectly sensible. We nodded in approval at the consultants. Great! We have a real schedule! We might actually pull this off! Until we realized that:

1. Nobody can accurately estimate their time or all the steps they'll need to perform.
2. Trying to get into that much detail that far out is useless. Something will always spoil your plan.
3. We were spending all our time scheduling, arguing over what could and couldn't be done in a half day, and it was impossible to see the whole forest through the half trees.

Whenever the product would shift and evolve, we'd scramble. We'd have to harass everybody to tell us how many half days it would take them to deal with the change, instead of just dealing with it. We'd spend hours every week to "work the schedule" with each member of the team instead of actually working.

After a few months, we scrapped the whole system. No more half days. We organized our time into bigger chunks—weeks, months. We started taking a macro view of our projects. And that enabled us to build the V1 of Velo in about eighteen months. Then we handed it, gleaming and new, to sales and marketing.

And they had absolutely no idea what to do with it. They'd never seen it before. They didn't know how to sell it, where to sell it, how to advertise it. They had been an afterthought to us, and now we were an afterthought to them.

We had figured out our internal heartbeat, but had never synced it with any other team. Nobody else could keep up with our rhythm. We were dancing to our own beat, sure that all eyes were on us, and our dance partner was across the room getting punch, thinking about electric shavers.

We needed internal milestones within the project—regular check-ins where we would make sure everybody understood how the product had evolved and could evolve their side of the business along with it. And to make sure the product still made sense. To see if marketing still liked it. To see if sales still liked it. To see if support could still explain it. To make sure everyone knew what they were making and the plan to launch it.

Those milestones slow you down in the short term, but ultimately speed up all of product development. And they make for a better product.

And then, eventually, finally, one day you'll be done. Or at least done enough. And you'll reach your very first external heartbeat for V1.

Hopefully it goes well. Hopefully the world likes it. Hopefully they want more, so that first external heartbeat will be followed by another. And another.

Once you move past V1 and onto V2, the pace of your external announcements, and possibly your competitors, will begin to drive your internal heartbeats.

Just be careful.

If you're building something digital—an app, a website, a piece of software—you can literally change your product at any time. You can add features every week. You can redesign the whole experience once a month. But just because you can, doesn't mean you should.

Heartbeats shouldn't be too fast. If a team is constantly updating their product, then customers start tuning out. They don't have time to learn how the product works—certainly not to master it—before suddenly it's new again.

Look at Google. Its heartbeat is erratic, unpredictable. It works for them—mostly, sometimes—but it could work so much better. Google arguably only has one big external heartbeat each year at Google I/O—and most teams don't bother aligning with it. They typically launch whatever they want whenever they want throughout the year, sometimes with real marketing behind it, other times with simple email campaigns.

That means they can never communicate with their customers in a cohesive way about their entire organization. One team does this, another does that, their announcements either overlap or ignore obvious opportunities to create a narrative. And nobody, not customers, not even employees, can keep up.

You need natural pauses so people can catch up to you—so customers and reviewers can give you feedback that you can then integrate into the next version. And so your team can understand what the customer doesn't.

But you also can't slow down too much. The heartbeats of companies that work with atoms rather than electrons are often way too slow. Because atoms are scary: you can't relaunch an atom.

The right process and timing is a balancing act—not too fast, not too slow.

So look at the year ahead.

After you've launched your V1, then two to four times in that year, you should be announcing something to the world. New products, new features, new redesigns or updates. Something meaty that's worth people's attention. It doesn't matter if you're a big or tiny company; if you're building hardware or apps, B2B or B2C, this is the right rhythm for customers. For humans. Any more announcements or big changes and you'll start confusing people, any fewer and they'll start forgetting about you. So have at least one really big launch and another one to three smaller launches every year.

Apple's external heartbeat used to thump loudest at the annual MacWorld conference in San Francisco. That event drove the pace of the whole company. The biggest announcements always had to happen at MacWorld.

And MacWorld always happened in January.

The main reason was that MacWorld organizers were cheap. The first week of the year was the least expensive time to rent conference space in San Francisco since tourists and businesspeople took a break from traveling after the holiday rush. And in any case, MacWorld was small. In the nineties Apple was floundering and its customer base was tiny, so the few die-hards who attended the show were Silicon Valley techies who were already in the neighborhood. The city of San Francisco was happy to keep the geeks coming in January and save more lucrative spring and summer slots for larger conferences that would attract more out-of-towners.

So January it was.

But that meant that every year, Apple couldn't take time off over the holidays. Absolutely everything had to be done by January 1. If you worked on certain teams at Apple, your family just resigned themselves to never seeing you from Thanksgiving until the new year. Most teams would only reemerge after MacWorld, haggard but triumphant, squinting and rubbing their eyes in the sunlight. This went on for years and years and years.

Until, eventually, Steve Jobs said, "Screw this."

He decided Apple was strong enough to skip MacWorld. He set a new heartbeat.

The old heartbeat had big announcements at MacWorld in January and smaller releases at the Apple Worldwide Developers Conference (WWDC) in June and then again in September.

But the new heartbeat was smaller announcements in March, then a big blowout at WWDC in the summer, and more smaller announcements in the fall.

Now Apple has so much to talk about that there are announcements in March, June, September, and October, just before the holidays.

But not January. Never January. They learned that lesson well.

Unfortunately, you don't always control your heartbeat. Sometimes it's based around someone else's conference. Sometimes it orbits someone else's products.

For a long time Macintoshes were at the whim of IBM, Motorola, and Intel—the makers of their processors. If a new processor was delayed, then the new Macs would be delayed. That's why Macintosh settled on Intel processors for so long—because they were the least unreliable of the lot. But even Intel wasn't 100 percent predictable, and any slight change in their schedule caused endless scrambling and readjusting on the Apple side.

There was no way to create a steady heartbeat for Mac customers or a reasonable rhythm for the Apple team if they relied on Intel processors. So just as Steve decided to own his schedule for announcements, eventually he decided Apple needed to make its own processors.

That was the only way to make the world predictable.

And there's nothing people like more than a predictable world.

We like to think that we're not ruled by schedules, that we can throw off the chains of habit at any time—but most people are creatures of routine. They're comforted by the knowledge of what comes next. They need it to plan their lives and their projects.

Predictability allows your team to know when they should be heads down working and when they should be looking up to check in with other teams or to make sure that they're still headed in the right direction. [See also: Chapter 1.4: Don't (Only) Look Down.]

Predictability allows you to codify a product development process rather than starting from scratch every time. It allows you to create a living document with checkpoints, milestones, schedules, and plans that trains new employees and teaches everyone: This is how we do it. This is the framework for how to build a product.

Ultimately, that predictability is how you'll actually make your deadline.

Breaking the rhythm of your external heartbeat should be avoided at all costs, but sometimes it'll happen anyway. Something will break. Something will take longer than anyone expected. It almost always happens with V1, when you're starting from scratch, trying to figure out everything at once.

But once you've got your process in place and can finally get V1 out the door, your heartbeat can settle down. It can get steady.

And when you ship V2, you'll actually be on time. And everyone—your team, your customers, the press—will feel the rhythm.

Chapter

3.6

THREE GENERATIONS

The joke is that it takes twenty years to make an overnight success. In business, it's more like six to ten. It always takes longer than you think to find product/market fit, to get your customers' attention, to build a complete solution, and then to make money. You typically need to create at least three generations of any new, disruptive product before you get it right and turn a profit. This is true for B2B and B2C, for companies that build with atoms or electrons or both, for brand-new startups and brand-new products.

Keep in mind there are three stages of profitability:

1. **Not remotely profitable:** With the first version of a product you're still testing out the market, testing out the product, trying to find your customers. Many products and companies die at this stage before they ever make a dime.

2. **Making unit economics or gross margins:** Hopefully with V2 you can make a gross profit with each product sold or each customer who subscribes to your service. Keep in mind that fantastic unit economics are not enough to make a profitable company. You'll still be spending a ton of money just running your business and acquiring customers through sales and marketing.

3. **Making business economics or net margins:** With V3 you have the potential to make net profits with each subscription or product

sold. That means that what you take in in sales revenue outstrips your business costs, so your company as a whole makes money.

The reason it takes so long to reach gross margins and even longer to make net margins is that learning takes time. For your company and your customers.

Your team will have to figure out how to find product/market fit for V1, then get the product fixed up and properly marketed to a wider audience with V2, and only then can you focus on optimizing the business so it can be sustainable and profitable with V3.

And customers need time to feel you out. The vast majority of people aren't early adopters—they won't try new things right away. They need time to get used to the idea, time to read some reviews, time to ask their friends, and then time to wait until the next version comes out because that'll probably be even cooler.

••

I remember walking around the halls of General Magic reading *Crossing the Chasm* by Geoffrey Moore back in 1992 or 1993. A lot of us were reading it, discussing it, pointing out how right it was even as we hurtled deeper and deeper into the chasm and it became obvious we were never going to make it out.

The "chasm" is the hole companies can fall into if regular people—not just early adopters—won't buy their product. Today it's called finding product/market fit.

Crossing the Chasm introduced the world to the famous Customer Adoption Curve chart below. The idea behind it is pretty simple: a small percentage of customers will jump to buy a new product early regardless of how well it works—they just want the latest doohickey. However, most will wait until it's been around for a while and all the kinks have been worked out.

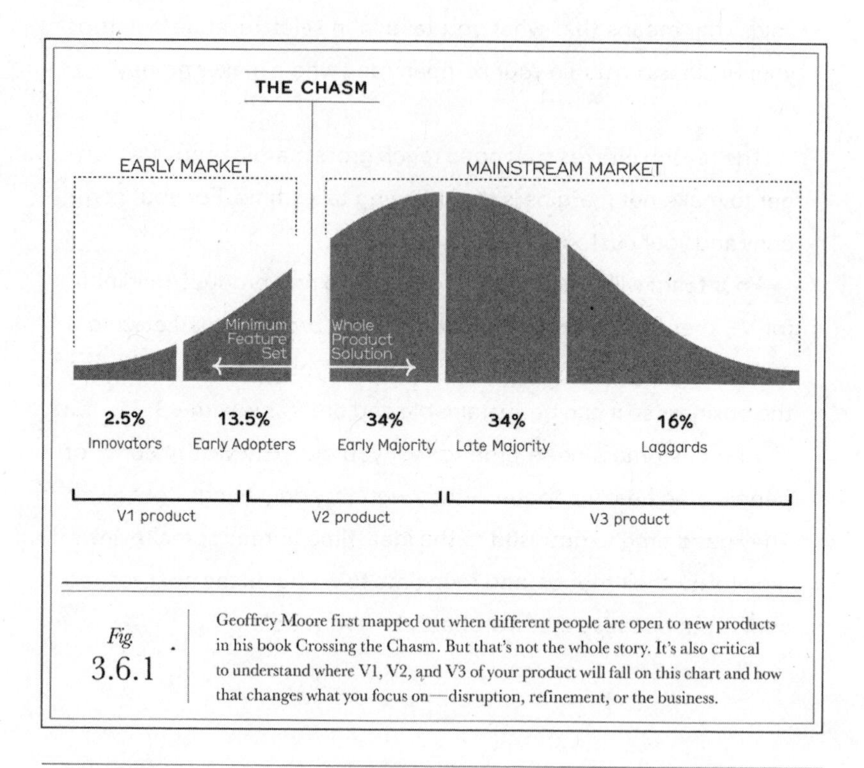

Fig. 3.6.1 · Geoffrey Moore first mapped out when different people are open to new products in his book Crossing the Chasm. But that's not the whole story. It's also critical to understand where V1, V2, and V3 of your product will fall on this chart and how that changes what you focus on—disruption, refinement, or the business.

Who's it for

V1

Innovators and early adopters.

These are the people who deeply love whatever it is you do—they might be gear heads or obsessed with technology or just really into your space. They'll have an emotional reaction to anything new and cool and will buy it fully aware that it will probably be buggy.

V2

Early majority.

These are the trendsetters. They watch the early adopters and will read some reviews before committing. They expect product bugs to be ironed out, decent customer support, and an easy way to learn about and purchase the product.

V3

Late majority and laggards.

This is everyone else—the mass customers who expect perfection. They'll only buy the clear winner in the marketplace and won't put up with any hassles.

Product

V1	V2	V3
You're essentially shipping your prototype.	You're fixing the stuff you screwed up with V1.	You're refining an already great product.
Customer acquisition costs will be sky-high; some features you really wanted will be missing; your marketing, sales, and customer support will all be a little wonky; you won't have the partnerships you'll need; and you'll still be discovering everything you got wrong.	At this stage you'll know what your problems are and how to fix them—both the unexpected issues that inevitably crop up after you launch and the stuff you cut corners on the first time. V2 usually comes swiftly after V1 because you've learned so much so fast and you're dying to get it all into the next generation.	Your focus should be less on the product and more on the business and polishing every touchpoint of the customer lifecycle.

Outsourcing vs. building in-house

V1	V2	V3
Figuring things out and outsourcing.	Start bringing more things in-house.	Lock in internal expertise and selectively outsource smaller projects.
You have a small team so you have to outsource a ton of functions—marketing, PR, HR, legal. This will enable you to move fast and get a lot done, but it's expensive and won't scale.	You take everything you've learned from the third-party teams you worked with for V1 and start building that muscle in-house. [See also: Chapter 5.3: Design for Everyone: But you shouldn't outsource a problem.] Your teams and level of expertise grow.	Certain key internal teams will be focused on the most important differentiators of your business. That may mean doing branding in-house or legal or whatever is most important to your company. As these teams grow and do more, they begin outsourcing again, but only for specific, smaller tasks that are closely supervised by the internal team.

Product

V1	V2	V3
Product market fit.	Profitable product.	Profitable business.
This really just means getting the product right enough to prove there's a market for it so you can cross the chasm. If you can't prove that at least early adopters will buy your V1, then you have to go back to the drawing board and start over.	At this point you'll widen the market, start nailing more parts of the customer journey, and might even be making a little money per product—but probably not enough to cover costs.	Assuming you reached gross margins with V2, you may want to aim for net margins with V3. This is when you begin negotiating with partners for better deals, optimizing your customer support and sales channel, and buying new kinds of media for marketing. Hopefully you'll finally get enough volume that you can start to reduce your prices and actually make real money. With V3 you have a chance to get it all right—the product, the company, and your business model.

But that's not the whole story. If you don't understand how customer adoption maps onto product and company development, you're missing a very important part of the puzzle.

After companies find product/market fit, they can start to focus on profitability. Businesses that build with atoms are focused on COGS—cost of goods sold. Aside from direct labor, the main thing they spend money on is actually making the product. So they need to lower the cost of producing their product in order to reach profitability.

Companies that build with electrons are focused on CAC—customer acquisition costs. Aside from direct labor, their money gets spent selling and supporting their product.

Companies that build with both atoms and electrons have to worry about COGS and CAC, but generally should focus on one at a time. First knock out COGS, then move on to CAC. Build the product, then add the services.

And despite the many differences between atoms and electrons, hardware and software, there is one thing that has the exact same stranglehold on both: time.

No matter what you're building, reaching profitability will take longer than you think. You will almost certainly not make any money with V1. You'll need to reinvent yourself at least three times. Sometimes many more.

And even if your timeline has shrunk—even if you're just revving an app—your product still has to learn how to crawl, and then walk, before it can run. That can take just as long for an app or service as it does for a hardware launch. It takes time to evolve and change, to react to customer feedback, to make every point on the customer journey as strong as the product itself. And customers still need time to learn about you, to try your product, to decide it's worth it. They need time to march up the adoption curve.

The iPod took three generations—and three years—before it reached profitable unit economics.

Same story with the iPhone. The first version was really only for early adopters—it didn't have 3G, didn't have the app store, and our pricing model was all wrong. Steve never wanted the phone to be subsidized—he wanted everyone to know its real price so they could value it appropriately—and he wanted to get a cut of the data plan, too. [See also: Chapter 6.4: Fuck Massages: When people pay for something, they value it.] But the iPhone was destined to cross the chasm—the world loved it. They just needed us to get the details right before they bought it.

But crossing the chasm isn't a guarantee, even with much-loved products. And actually making money is much, much harder.

Of course with the internet, new business models challenge this conventional wisdom. Even so, many companies—Instagram, WhatsApp, YouTube, Uber—have gone through five or ten or more generations before they figured out how to make money. Many others still haven't. The reason unprofitable companies are still around is that they have a giant pool of VC funding or were acquired by even larger tech companies. They focused on finding product/market fit and building their user base first, and figured they'd iterate on the business model to make money later. But that does not work for everyone. It relies on a swift dive across the chasm and then a long, meandering doggy paddle to profitability through a huge pool of capital. That can doom a company just as fatally as falling into the chasm on the first step.

A few years ago the major cities of the world were flooded with scooter- and bike-sharing companies. All at once it seemed like they were everywhere. And that was the approach—these companies wanted to get as much market share as they could in order to acquire customers.

They had enough capital that they just bought up whatever bikes they could and expanded and expanded and expanded.

But they could never make it profitable. They couldn't get to V2 or V3. By the time they started figuring it out, they ran out of money—the endless pool ran dry.

Now the second- and third-generation scooter and bike companies are emerging, but they're taking a completely different approach after watching their predecessors flame out. They're extremely selective about their markets and are choosing the right atoms—incredibly durable bikes and scooters. They're being wary about where they spend their money and making sure they understand the unit economics in excruciating detail.

Having that kind of laser focus on a few key differentiating elements is a much likelier way to reach your goals than throwing out a wide net and hoping for the best.

In the early days, Tesla was so focused on the car itself—and really only several parts of the car—that almost nothing else mattered. They had basically no customer support—there was nobody you could talk to on the phone. So if your Tesla had an issue, they'd just come to your house and drive it away. You'd be left without a car, wondering what you were supposed to do next.

Luckily Silicon Valley, Tesla's home base, is filled with many tech enthusiasts and early adopters. A friend of mine bought one of the earliest Tesla Roadsters—their V1. It was really an electrified Lotus, hadn't been fully redesigned from the ground up, but it had one of Tesla's core features: regenerative brakes. Every time you hit the brakes, your car would use the motor as a generator to charge up the battery.

The trouble was my friend lived on a mountaintop. So he'd drive up the mountain and plug in his car for the night, but when he drove back down in the morning, his brakes barely worked. Turned out he couldn't charge the Tesla to 100 percent—keeping his foot on the brake on the way down the mountain overcharged the battery. Tesla had to fix their braking and charging algorithms to keep him from crashing.

But my friend was the definition of an early adopter—he loved his Roadster. Even though it spent more time in the shop than his driveway. Even when he started just calling the engineers directly when he had an issue.

Early adopters know nobody gets everything right with V1. Nobody even gets everything they were originally planning for V1 into V1. The product and customer base evolve and grow with each iteration, and every stage brings on different risks and challenges and investments. Nobody can tackle them all at once. Not at a startup, not at a big company.

So you and your employees and your customers all need to have the right expectations. And so do your investors.

Far too many people expect profitability, for the product and the business, right out of the gate. When I was at Philips I watched most new product categories and businesses on their slate get cancelled—even for products that were almost ready to ship. Built. Tested. Done. They would die on the vine because the top brass were protecting themselves. Any execs joining the team always wanted a near guarantee that new products would make money. [See also: Chapter 2.2: Data Versus Opinion: Most people don't even want to acknowledge that there are opinion-driven decisions.] They demanded to be shown ahead of time that the unit and business economics of the product were sound. But that was impossible.

They were asking us to predict the future with near 100 percent confidence. They were asking for proof that a baby could run a marathon before it had even learned to walk.

These guys didn't know much about babies. They knew even less about how to create a new business.

That's why so many Kickstarter projects have failed. They thought, "If I build it for $50 and sell it for $200, then I'll make money. My company will be a success." But that's not how companies work. That $150 profit gets sucked away with every new office chair and dependent on your employees' insurance, with every customer support call and Instagram ad. Until you optimize the business, not just the product, you can never build something lasting.

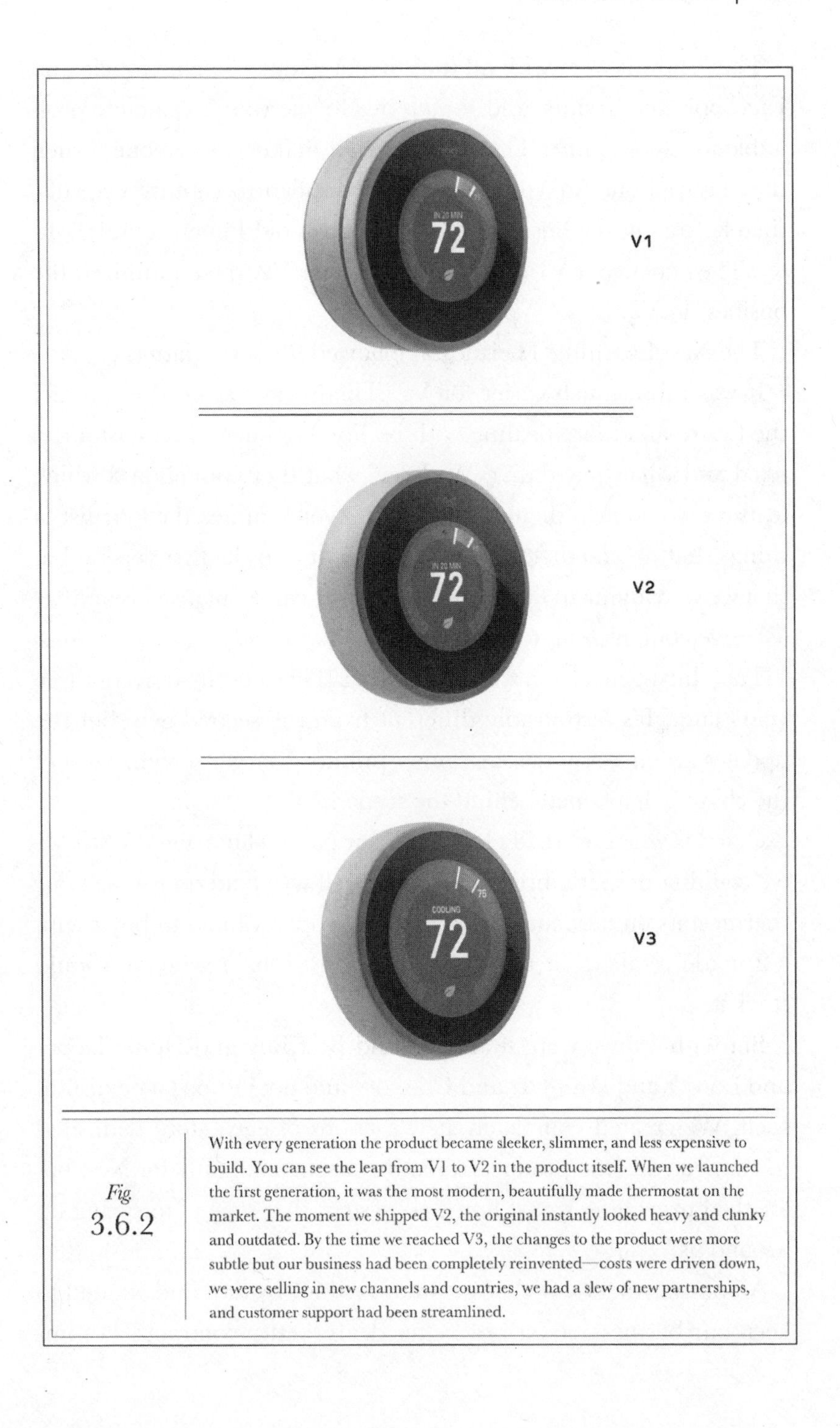

Fig. 3.6.2

With every generation the product became sleeker, slimmer, and less expensive to build. You can see the leap from V1 to V2 in the product itself. When we launched the first generation, it was the most modern, beautifully made thermostat on the market. The moment we shipped V2, the original instantly looked heavy and clunky and outdated. By the time we reached V3, the changes to the product were more subtle but our business had been completely reinvented—costs were driven down, we were selling in new channels and countries, we had a slew of new partnerships, and customer support had been streamlined.

That's how it went with all the current giants of tech: Google and Facebook and Twitter and Pinterest. Google wasn't remotely profitable for a long time. They only started making real money when they figured out AdWords. Facebook decided to capture eyeballs, then figure out the business model later. So did Pinterest and Twitter. They created a V1 product, scaled it for V2, then optimized the business in V3.

The Nest Learning Thermostat followed the same pattern.

It was all so much easier for V2. There was less need to predict the future and more dealing with reality. We knew what customers liked and what they didn't. We knew what they wanted and which features would help them most. And we got to attack the long list of things that we couldn't quite manage to fit into the first version but that we were dying to fix for the second. V2 came out just a year after V1—we couldn't wait to launch it.

The third-generation Nest Learning Thermostat arrived three years later. It's also visibly different from the second gen, but the updates are more subtle. A slimmer profile. A larger screen. Most of the changes happened behind the scenes.

Gen 3 is when we really locked in our channel partners. With V1 we couldn't properly break into retail. All we could do was sell our thermostats on nest.com to prove that people wanted to buy them. V2 made retailers sit up and consider us: Oh, maybe we should stock it.

But with V3 we were in Target and Best Buy and Home Depot and Lowe's and Walmart and Costco—and not just on some distant shelf. We created completely new sections of each store dedicated to connected home products, creating a space not only for Nest but for the burgeoning smart home ecosystem that started to spring up around us.

All our partners saw that we were getting traction and wanted to keep our business, so we got better deals, better contracts. We improved customer support, got our price per call way down, and fixed up our knowledge base.

So when we started work on the Nest Protect smoke and CO alarm, our second product, you'd think it would be easier. That everything we'd built already would let us skip a few steps. But the second you start a new product, you have to hit the restart button—even if you're at a big company. Sometimes it's even harder the second time around because all the infrastructure that's been built up for the first product gets in the way. So you'll still need to go through at least three generations before you get it right.

You make the product. You fix the product. You build the business.

You make the product. You fix the product. You build the business.

You make the product. You fix the product. You build the business.

Every product. Every company. Every time.

Part

IV

BUILD YOUR BUSINESS

I have to start this company, don't I?

Shit.

A startup was not the plan. The plan was a break. A long one. I needed it—I'd finally left Apple in 2010 after almost a decade of heads-down sprinting. We had launched the first three generations of the iPhone and now the big swings were over. After eighteen generations of iPod, I knew how this would go—we'd make tweak after tweak after tweak ad infinitum. Or I could start on the iPad, which had basically the same guts as the iPod Touch, which was basically an iPhone.

But the bigger reason I quit was my family. I'd met my wife at Apple—Dani was vice president of HR. We had two very young children. And although we always made time for them, we also worked a ton. This was a chance to live differently. So Dani and I both left Apple. Then we left the country.

We traveled all over the globe and worked hard not to think about work. But no matter where we went, we could not escape one thing: the goddamn thermostat. The infuriating, inaccurate, energy-hogging, thoughtlessly stupid, impossible-to-program, always-too-hot-or-too-cold-in-some-part-of-the-house thermostat.

Someone needed to fix it. And eventually I realized that someone was going to be me.

The big companies weren't going to do it. Honeywell and the other white-box competitors hadn't truly innovated in thirty years. It was a dead, unloved market with less than $1 billion in total annual sales in the United States. And after a failed green innovation wave in 2007 and 2008, green-tech investors had firmly turned away from energy-saving devices. A small startup filled with fresh faces and few connections wouldn't have the credibility to get funding. I could already hear the VCs sneer, "Thermostats? Really? You want to make thermostats? The market is tiny and boring and hard."

But one day I was on a bike ride with Randy Komisar. Randy is a longtime friend, mentor, and a partner at the venerable venture firm Kleiner Perkins. We first met when I pitched him on investing in Fuse in 1999. Since I deeply trusted Randy, I decided to test my idea on him and floated the idea of a smart thermostat.

He offered to write me a check on the spot.

I was exactly the kind of founder investors like. Four failed startups and years of professional disappointment had paved the way for a decade of success. I was forty years old, knew exactly how hard this was going to be and which mistakes not to make again. I'd worked on hardware and software at tiny and enormous companies. I had contacts, credibility, and enough experience to know what I didn't know. And I had an idea.

Your thermostat should learn the temperatures you like and when you like them. It should connect to your smartphone so you can control it from anywhere. It should turn itself down when you're not home to save energy. And of course it should be beautiful—something you'd be proud to put on your wall.

The only thing missing was the will to take the plunge. I wasn't ready to carry another startup on my back. Not then. Not alone.

Then, magically, Matt Rogers reached out to me. Matt had started as one of the first interns early on the iPod project and I'd watched him rocket ahead of everyone else on an already world-class hand-picked team. We hired him full-time after graduation and he quickly became a fantastic manager—focused on team building, never afraid to ask questions or push boundaries, insatiably curious about every aspect of the business.

After I left Apple, he started getting frustrated with how things were going. So we had lunch and he asked what I was going to do next. I told him the idea. He was enthusiastic. And when I say enthusiastic, you should know that Matt is a perpetual motion machine of unstoppable energy. He immediately started digging in, offering suggestions and ideas, growing more and more excited the longer we talked.

That push was what I needed to commit. He was a real partner who could share the load, who'd work just as hard as me and care just as much. We already knew how to work together and saw eye-to-eye on how to make products. I didn't need another middle-aged executive with decades of experience telling me what we couldn't do. I needed a real cofounder. I needed Matt.

Together we took the idea and formed it into a vision. What we pitched to investors was a connected thermostat. But we knew the company we were really building—it wasn't going to stop at thermostats. We would create a slew of products, reinventing unloved but important objects everyone needed at home. And, most importantly, we were going to create a platform. We were going to build the connected home.

The concept wasn't new. Connected home systems had been around for a couple of decades by then. I remember Bill Atkinson from General Magic trying to build a connected home in the nineties. He tried to rig it himself, battling to make something useful. But over the years a lot of wealthy tech people just dropped a quarter million dollars to have an elaborate system built into their walls. There'd be sensors and screens and switches and controllers for thermostats, alarm systems, lights, music. All very shiny, very fancy. And complete crap. Garbage. None of it worked.

Our investors winced when we mentioned it in our pitch. Yes, they'd all been taken in. Yes, all their spouses were still pissed about it.

We wanted to take a different approach. Instead of trying to cram a fully formed platform with every possible gadget into someone's house, we were going to start with just one really good product—a beautiful thermostat that would live on people's walls for a decade or more. Once people fell in love with our thermostat, they'd buy more products that worked with it. Customers would be able to assemble a connected home piece by piece, creating a unique system that made sense for their house and their family.

The thermostat would be our way in.

But first we had to build the thermostat.

Making it beautiful wasn't going to be hard. Gorgeous hardware, an intuitive interface—that we could do. We'd honed those skills at Apple. To make this product successful—and meaningful—we needed to solve two big problems:

It needed to save energy. And we needed to sell it.

In North America and Europe, thermostats control half a home's energy bill—something like $2,500/year. Every previous attempt to reduce that number—by thermostat manufacturers, by energy companies, by government bodies—had failed miserably for a host of different reasons. We had to do it for real, while keeping it dead simple for customers.

Then we needed to sell it. Almost all thermostats at that point were sold and installed by professional HVAC technicians. We were never going to break into that old boys' club. We had to find a way into people's minds first, then their homes. And we had to make our thermostat so easy to install that literally anyone could do it themselves.

So we got to work.

Fig 4.0.1 The Nest Learning Thermostat launched in October 2011 for $249. It had a unique round 2.75-inch screen and measured 3.2 x 3.2 x 1.6 inches. It came with its own mobile app and had a built-in AI that learned your schedule and turned itself down when you were away.

It was fatter than we wanted. The screen wasn't quite what I imagined. Kind of like the first iPod, actually. But it worked. It connected to your phone. You could install it yourself. It learned what temperatures you liked. It turned itself down when nobody was home. It saved energy.

And people loved it.

Before we launched we didn't know if anyone would be interested, so we didn't want to spend all our money and have a ton of inventory sitting in warehouses. Incredibly, we sold out on the first day, then we were continually sold out for more than two years.

We quickly followed up with the second-generation thermostat, fixing all the things we hadn't gotten quite right with the first one.

And then we focused on our next product. What device existed in every single home and was even more infuriating than the thermostat?

Easy: the smoke alarm.

The annoying, false-alarming, constantly-going-off-while-you're-cooking-and-chirping-at-2-a.m.-so-you-have-to-hunt-down-which-stupid-alarm-has-low-batteries-and-of-course-it-turns-out-to-be-the-one-you-can't-reach smoke alarm.

Had we realized how hard it would be to innovate in the smoke and CO space, we probably wouldn't have even started. But all we knew was that smoke alarms were everywhere, in every room of every house. And they were awful. Truly, monumentally terrible. They were required by law in every home, so smoke alarm manufacturers had no incentive to make them any better—awful or not, they had to be everywhere.

But their awfulness was so all-encompassing that people would literally risk their lives to make them shut up—removing batteries or tearing smoke detectors off the wall after one too many false alarms, knocking them off the ceiling with a golf club in the middle of the night just to stop the infernal chirping.

So in 2013 the Nest Protect smoke and CO alarm was born.

Fig. 4.0.2 The Nest Protect retailed for $119, was 5.28 x 5.28 inches, and offered protection from carbon monoxide and smoke. False alarms could be hushed from the app and you'd get an alert on your phone if there was danger.

I had made successful products before—the iPod, the iPhone—but Nest was my first time trying to truly build a large, successful business. It was the first time I started from scratch—from a single cell of an idea—and watched that cell divide and grow into a fully formed baby. Our baby. Our company.

So if you want to start a business or a new product or project within a big company—or if you've already started and are looking on in joy, fear, and amazement as it begins to take on a life of its own—then here's what I've learned about choosing an idea, starting a company, finding investors, and almost passing out from stress.

This is what I've figured out so far about every stage of growth—and what you should do when your baby isn't a baby anymore.

Chapter

4.1

HOW TO SPOT A GREAT IDEA

There are three elements to every great idea:

1. It solves for "why." Long before you figure out what a product will do, you need to understand why people will want it. The "why" drives the "what." [See also: Chapter 3.2: Why Storytelling.]

2. It solves a problem that a lot of people have in their daily lives.

3. It follows you around. Even after you research and learn about it and try it out and realize how hard it'll be to get it right, you can't stop thinking about it.

Before you commit to executing on an idea—to starting a company or launching a new product—you should commit to researching it and trying it out first. Practice delayed intuition. This is a phrase coined by the brilliant, Nobel Prize–winning economist and psychologist Daniel Kahneman to describe the simple concept that to make better decisions, you need to slow down.

The more amazing an idea seems—the more it tugs at your gut, blinds you to everything else—the longer you should wait, prototype it, and gather as much information about it as possible before committing. If this idea is going to eat up years of your life, you should at least take a few months to research it, build out detailed (enough) business and product development plans, and see if you're still excited about it. See if it will chase you.

And keep in mind that not all decisions rise to this level. Most day-to-day decisions can and should be made quickly, especially if you're iterating on something that already exists. You should still take your time and consider your options and make sure you're thinking through next steps, but not every idea has to chase you around for a month.

••

The best ideas are painkillers, not vitamins.

Vitamin pills are good for you, but they're not essential. You can skip your morning vitamin for a day, a month, a lifetime and never notice the difference.

But you'll notice real quick if you forget a painkiller.

Painkillers eliminate something that's constantly bothering you. A regular irritation you can't get rid of. And the best pain—so to speak—is one you experience in your own life. Most startups are born from people getting so frustrated with something in their daily experience that they start digging in and trying to find a solution.

Not every product idea has to come from your own life, but the "why" always has to be crisp and easy to articulate. You have to be able to easily, clearly, persuasively explain why people will need it. That's the only way to understand what features it should have, whether the timing is right for it to exist, whether the market for it will be tiny or enormous.

Once you have a really strong "why," you have the germ of a great idea. But you can't build a business on a germ. First you have to figure out if this idea is actually strong enough to carry a company. You need to build a business and implementation plan. And you have to understand if it's something you want to work on for the next five to ten years of your life.

The only way to know is to see if it will chase you. And the chasing process is always the same:

- First, you're dumbstruck by how great an idea this is. How has nobody thought of this before?
- Then you start looking into it. And oh, okay—they have thought of it. They tried it and failed. Or maybe you really have stumbled into something nobody has ever done before. And the reason is this insane, impossible obstacle that there's no way to get around. You begin to understand how hard it would be to do this—there's so much you don't know. So you put it aside.
- But you can't get it out of your head. So you research it here and there. You start sketching or coding or writing, making little prototypes of what this thing could be. Napkin drawings constantly fall out of your bag. Your notebook is full of feature ideas, sales ideas, marketing ideas, business model ideas. You think that maybe the people who tried this idea before were taking the wrong approach. Or maybe an obstacle that was stopping others can now be solved with a new technology—maybe the moment for this idea has finally come.
- That's when it starts to get more real for you. So you decide to commit to really looking into it, to digging deep to make an informed decision. You need to figure out whether you should pursue this idea or not.
- One day you realize there's a way around that one impossible obstacle. You're thrilled! Until you see the next huge roadblock in your way. Crap. It'll never work. But you keep digging and trying things and getting advice from experts and friends and you realize, actually, maybe there's a way around that, too.
- People start asking you about your project—when are you going to start? Can I join? Are you taking angel investment? Each obstacle turns into an opportunity, each problem pushes you to find a new solution, and each solution gets you more excited about this idea.

- Even though there are still a million unknowns—they're not unknown unknowns anymore. You understand the space. You can see what this business could become. And you start to gain momentum from all the research you've done and barriers you've conquered. It feels like it's all coming together. You feel in your gut that this is the right decision. So you bite the bullet and commit.

For me, this whole process took ten years. That's how long the thermostat chased me.

That is pretty extreme, by the way. If you've got an idea for a business or a new product, you usually don't have to wait a decade to make sure it's worth doing.

But you should commit a month—or two, or six—to researching and poking around and making some rough prototypes, articulating your story of "why." If during that month—or two, or six—you only get more excited about the idea and can't stop thinking about it, then you can get more serious. Take at least another few months—possibly up to a year—to look at it from every angle and consult with people you trust, create some business plans and presentations, and prepare as well as you possibly can.

You do not want to start a company only to discover that your seemingly great idea is a shiny veneer over a hollow tooth, ready to crack at the slightest pressure.

A lot of startups have a "fail fast" mentality in Silicon Valley. It's a trendy term that means instead of planning carefully for what you want to make, you build first and figure it out later. You iterate until you "find" success. This can manifest in two ways—either you knock out a product quickly then iterate even faster to get to something people want, or you quit your job, cut loose from your commitments, and sit around thinking up startup ideas until you find a business that works. The former approach sometimes works; the latter usually fails.

Throwing darts at a wall is not how you pick a great idea. Anything worth doing takes time. Time to understand. Time to prepare. Time to get it right. You can fast-track a lot of things and skimp on others, but you cannot cheat time.

That said, ten years is a little much. But for most of the ten years that I idly thought about thermostats, I had no intention of building one. I was at Apple making the first iPhone, leading a huge team. I was learning and growing, up to my eyeballs in work. And eventually I got married, had kids. I was busy.

But then again, I was also really cold. Bone-chillingly cold.

Every time my wife and I drove up to our Lake Tahoe ski cabin on Friday nights after work, we'd have to keep our snow jackets on until the next day. The house took all night to heat up since we kept it just above freezing when we were away to avoid wasting energy and money.

And I felt the pain. Walking into that frigid house drove me nuts. It was mind-boggling that there wasn't a way to warm it up before we got there. I spent dozens of hours and thousands of dollars trying to hack security and computer equipment tied to an analog phone so I could fire up the thermostat remotely. Half my vacations were spent elbow-deep in wiring, electronics littering the floor. My wife rolled her eyes—you are on vacation! But nothing worked. So the first night of every trip was always the same: we'd huddle on the ice block of a bed, under the freezing sheets, watching our breath turn into fog until the house finally warmed up by morning.

Then on Monday I'd go back to Apple and work on the first iPhone.

Eventually I realized I was making a perfect remote control for a thermostat. If I could just connect the HVAC system to my iPhone, I could control it from anywhere. But the technology that I needed to make it happen—reliable low-cost communications, cheap screens and processors—didn't exist yet. So I tried to set the idea aside. Focus on my work. Not think about the cold.

A year later we decided to build a new, super-efficient house in Tahoe. During the day I'd work on the iPhone, then I'd come home and pore over specs for our house, choosing finishes and materials and solar panels and, eventually, tackling the HVAC system. And once again, the thermostat came to haunt me. All the top-of-the-line thermostats were hideous beige boxes with bizarrely confusing user interfaces. They touted the fact that they had touchscreens and clocks and calendars and showed digital photos. None of them saved energy. None could be controlled remotely. And they cost around $400. The iPhone was selling for $499.

How did these ugly, piece-of-crap thermostats cost almost as much as Apple's most cutting-edge technology?

The architects and engineers on the Tahoe project heard me complaining over and over about how insane it was. I told them, "One day, I'm going to fix this—mark my words!" They all rolled their eyes—there goes Tony complaining again!

At first they were just idle words born of frustration. But then things started to change. The success of the iPhone drove down costs for the sophisticated components I couldn't get my hands on earlier. Suddenly high-quality connectors and screens and processors were being manufactured by the millions, cheaply, and could be repurposed for other technology.

And my life was changing, too. I quit Apple and began traveling the world with my family. And in every hotel room, every house, every country, every continent—every thermostat sucked. We were too hot, or too cold, or just couldn't figure out how to use them. The entire world had the same problem—this forgotten, unloved product that everyone had to have in their home was costing people tons of money on their bills and wasting untold amounts of energy while the planet just kept warming.

After that, the chase kicked into high gear. I couldn't shake the idea of making a connected thermostat. An actually smart thermostat. One that would solve my problem and save energy. One that would allow me to build on everything I had previously made.

So I let the idea catch me. I came back to Silicon Valley and got to work. I researched the technology, then the opportunity, the business, the competition, the people, the financing, the history. If I was going to upend my life and my family, take a huge risk, dedicate five to ten years to creating a device unlike anything I'd ever made in a space I knew nothing about, then I needed to give myself time to learn. I needed to sketch out designs. I needed to plan out features and think about the sales and business model.

During this time, I'd also play a little game with myself and people I truly respected. They'd ask me, "What's keeping you busy now? What's interesting to you?" So I'd tell them I had an idea—maybe a great idea—and share a few details to get their reactions and thoughts and questions. I was developing my pitch, figuring out the story of the product just like Steve would do. Then as the weeks of research and strategy began to come together, I stopped saying it was an idea and started saying I was building a product. Even though it wasn't quite true yet. But I wanted it to feel real—to get them, and especially me, to really dive to the details. I wanted to convince them and I wanted them to challenge it and I wanted to tell the story. To figure out if it held together.

It took around nine to twelve months of making prototypes and interactive models, building bits of software, talking to users and experts, and testing it with friends before Matt and I decided to take the plunge and actually pitch investors.

We didn't have perfect data that we'd succeed. No amount of research or delayed intuition will ever guarantee that. We probably had 40–50 percent of the risks of starting this company identified, with ideas for how to mitigate them. But there were still vast, yawning unknowns before us. In the end, even with all our hard work and preparation, this was an opinion-driven decision. [See also: Chapter

2.2: Data Versus Opinion.] So we went with our gut. It felt scary as hell, but it also felt right.

The interesting thing is that delayed intuition generally doesn't make it less scary. If anything, the more you understand it, the more butterflies in your stomach it'll give you. Because you'll uncover all the ways it can go wrong; you'll know the million things that might kill this idea and your business and your time.

But knowing what can kill you makes you stronger.

And knowing that you've already deflected some major bullets makes you stronger still.

That's why we didn't just present our vision when we pitched investors. We presented the why—told our story—and then we presented the risks. Too many startups don't know what they're getting into or, worse, try to cover up the risks of failure. But if investors can spot holes in your plan that you completely missed or ignored or carefully avoided, they won't have the confidence to fund you. So we listed our risks: building an AI, compatibility with hundreds of different (and ancient) HVAC systems, customer installation, retail, and the really big one—would anyone even care? Would the world want a smart thermostat? The potential company-destroying problems—and the steps to mitigate them—went on and on. But listing them out, breaking them down, talking honestly about them, that's what ultimately convinced investors that we really knew what we were getting into. And that we could make it work.

Eventually, each of those risks became a rallying cry for the team—instead of avoiding them, we embraced them. We continually said to ourselves, "If it were easy, everyone else would be doing it!" We were innovating. The risks and our ability to solve for them was what set us apart. We would do something nobody else thought possible.

That's ultimately what made this company worth starting.

Of course, all this is not to say that you should wait and research endlessly to make every little decision in your life. If you're not starting from scratch—if you're iterating—then everything speeds up.

It took me a decade to decide to build my first thermostat. Deciding to build a second version probably took a week. In fact, we already knew what the second version should be before we even finished the first one. We'd proven the market potential and the technology—now we just needed to refine it. Of course we'd make a second-generation thermostat. We'd already done the hard part. [See also: Chapter 3.4: Your First Adventure—and Your Second.]

If you're optimizing, you have data, constraints, and experience to lead you. You'll already know what it takes to get to V1, so reaching V2 won't be as much of a stretch. Or a mystery. V2 is never as scary as V1.

V1 is always completely, utterly terrifying. Always. Big, great, new ideas scare the living crap out of everyone who has them. That's one of the signs that they're great.

If you're reading this book, you're probably curious and engaged. And that means you will encounter many, many, many good ideas in your life. It seems like good ideas are everywhere. But the only way to know if they're truly great—meaningful, disruptive, important, worth your time—is to learn enough about them to see their huge potential risks, the vast downsides, the icy blue *Titanic*-sized shitshow that lurks just below the surface. At that point you'll probably set that idea aside. You'll move on to other opportunities, other jobs and journeys. Until you realize that no matter what you do, you can't stop thinking about that one idea. That's when you stop running from it and start chipping away at the risks, one by one, until you're confident enough that they're worth taking.

If that does not happen, then it's not a great idea. It's a distraction. Keep going until you find an idea that won't let you go.*

---

* If you're still struggling to decide whether to pursue an idea or not, I spoke more about this topic on the *Evolving for the Next Billion* podcast.

Chapter

4.2

ARE YOU READY?

The world is full of people who have an idea and want to start a company. Often they ask me if they're ready. Do I have what it takes to create a successful startup? Should I launch my project within a big company instead?

The answer is that you'll never know until you take the leap and try. But here's how to get as ready as you'll ever be:

1. Work at a startup.
2. Work at a big company.
3. Get a mentor to help you navigate it all.
4. Find a cofounder to balance you out and share the load.
5. Convince people to join you. Your founding team should be anchored by seed crystals—great people who bring in more great people.

••

The archetypal entrepreneur is a twenty-year-old kid who lucks into a brilliant idea in their mom's basement and watches it turn into a thriving company overnight. In the movie version, they parlay their technical genius into a flawed but effective leadership approach and watch the millions roll in. Then they buy a fancy car before learning the true value of friendship.

But that is not reality.

There's always an exception, an incredible wunderkind who rides a unicorn to the moon, but most successful entrepreneurs are in their late thirties and forties. There's a reason why investors prefer to back second-time entrepreneurs even if they failed the first time around. It's because these founders spent their twenties screwing up and learning. Most follow the same path I did: they work hard, fail and fail, take risks and go to doomed startups and try out giant companies and take the wrong job and luck into an amazing team and quit too early or don't quit soon enough. They bounce around like a pinball, constantly smacking their heads into something. They learn. Trial by fire.

According to the book *Super Founder,* by Ali Tamaseb, around 60 percent of the founders of billion-dollar startups started another company before their wild success and many lost a ton of money. Just 42 percent of them had a previous exit of $10 million or more, so the majority "failed" by the standards of venture capital.

But they came out on the other side with a basic mental model of a startup. They understood the operational details and what it might look like if that tiny startup became successful. That's it. That's the magical key to success.

The problem is that it takes years to get there. And everyone wants a shortcut.

But there's nothing that prepares you for starting a startup except working at a startup. So go get a job. Find a startup or small, nimble company with founders who know (more or less) what they're doing. You need a role model to mimic or an anti–role model to avoid. Stand in an office (or videoconference), watch it all come together, and get a feel for the basic building blocks:

What does an org chart look like?

What is sales?

How should marketing work?

What about HR, finance, legal?

You need a working knowledge of each discipline—not to be an expert in each, but to understand who you should hire, what their qualifications should be, where to find them, and when you'll need them. For example, you probably don't need HR at the start. You just need recruiting. You don't need finance, you need accounting. You can outsource legal for now, but what about creative? When do you need operations? When do you need customer support? And what kind? Customer support for a brick-and-mortar store looks very different from customer support for e-commerce.

Spend your time at your startup job understanding the business you're helping to build. And then go get another job—this time at a big company. That's the only way to get a handle on the problems and challenges that bigger companies face, especially those beyond the product—the organization, the processes, the governance, the politics. The more you can observe how each type of company operates, the fewer questions you'll have when you start your own.

Even if you have a brilliant idea for a world-changing product, when you're starting a business, you need to run that business. Making something new is hard enough—the unknown unknowns that keep you up at night should be focused on the problem you're trying to solve, not on whether to get a marketing agency or what kind of lawyer to hire. You won't have time to screw up the basics, to waste time learning the fundamentals.

Money burns fast. If you don't have the confidence to move forward quickly, you'll have to continually slow down to consult a hundred people about a thousand decisions. You'll get mired in options and opinions. "What's the best? What's the latest?" will rattle in your head endlessly. You'll lose sight of where you're trying to go in the face of all the different ways to get there.

Of course, that doesn't mean that you shouldn't consult with anyone. Doing it alone is impossible.

You're going to need a mentor or a coach.

You're going to need an incredible founding team.

And you'll probably need a cofounder.

Starting a company is unfathomably stressful and a truly ungodly amount of work and sacrifice. You need a partner who can balance you out, who you can call at 2 a.m. because you know they'll be awake, working on your startup, too. And who can call you when they're down and need support. This is going to be lonely and painful and exhilarating and exhausting and sharing the load is the only way to keep from being crushed by it.

But be careful—even if you have a cofounder, there can only be one CEO. And if you pile on the cofounders, you're asking for trouble. Having two founders works well. Three can work sometimes. I've never seen it work with more.

I remember one startup we worked with that had four cofounders. Every decision was made by consensus, which meant every decision took forever. They'd never started a company before, so even the most basic questions were endlessly debated—hiring, product changes, who to take money from, and how to structure the agreement. If they couldn't agree they would hem and haw, trying to be nice, trying to be reasonable, watering down their opinions until the company fell behind the competition, ran out of money, and the board had to swoop in, remove some founders, and change the whole team around.

Sharing the load is one thing; unloading it altogether is another. If you're going to lead a team, you need to be ready to lead.

When you close your eyes, you should already know exactly who your first employees will be. You should be able to write down a list of five names without a second thought. If you don't have that list of names ready before you start, you probably shouldn't be starting.

However, just having the list isn't enough. You actually need to hire them. At least a few. And having them commit, really commit, is very different from just hearing them say, "Yeah! That's cool. I'd love to work with you." If you can't get them to sign on the dotted line, you may need to rethink the whole thing.

Because in the beginning you're not going to have HR to help you find and hire a world-class team. You won't even have a recruiter. For the first twenty-five or so employees it'll all come down to you and your cofounder—your vision, your network, your ability to convince people that you know what you're doing. You can lean on your mentors and board (and hopefully early investors), you can put them to work to prop up your reputation, but ultimately you're selling yourself and your vision for success.

You need a story people can get behind. [See also: Chapter 3.2: Why Storytelling.] People you respect. People who will help you create something great. Your team is your company. And your first hires are crucial—they'll help you architect what your business and culture will become.

Every member of your founding team should be proven and great at what they do (consider any failed startups in their past a bonus—that means they know what to avoid this time around), but they also need to have the right mindset. Getting from 0 to 1 is a huge lift that asks a lot of everyone, especially considering it may not pay off. So you need individual contributors who will enthusiastically take the leap with you, either because they're as excited by the idea as you are, or because they're simply young or ambitious, or because they've already had some financial success and aren't worried about paying rent for a while.

Titles, pay, and perks should never be your main draw, but that doesn't mean you should be cheap. Try to be reasonably flexible and structure compensation so it fits the individuals you're hiring. Some people may prefer cash over equity and that should always be an option. But most of your team should get generous equity packages—

they are owners of the idea, too, so they should also be owners of the company. You want your team to have a vested interest in your success so when things go wrong—and they definitely will—those people will stick with you.

In those very, very early days you want people who are there for the mission above all. You're looking for passion, enthusiasm, and mindset. And you're looking for seed crystals.

Seed crystals are people who are so good and so well loved that they can almost single-handedly build large parts of your org. Typically they're experienced leaders, either managers of large teams or super-ICs who everyone listens to. Once they're in, a tidal wave of other awesome people will typically follow.

That's how we built our core team at Nest. We sought out the best of the best, and they created their own gravity—pulling in more and more talent.

I remember looking around the office in those early days with my mentor, the gritty, boisterous, wise Bill Campbell. We stood there just grinning.

I first met Bill when he was on the board of directors at Apple. I got in touch again when I needed help as I was starting Nest. I remember he looked directly into my eyes with a dead stare, watching for any microexpressions, and asked, "Are you coachable?" Which meant, "Will you listen? Are you ready to learn?" That was the only qualification you needed to be coached by Bill—the ability to admit that you don't know everything. That you're going to screw up. And that you're ready to learn from those screwups, listen to advice, and act on it.

Bill wasn't remotely technical, had never been an engineer, but he knew people. He knew how to work with them and get the best out of them. He could tell me how to run a board meeting. He could tell me what to do if my team got stuck. And he could always see issues coming a mile away. When he saw I was about to take a wrong turn, he'd put his finger in his mouth, make a popping noise, and say, "You know what that is? That's the sound of you pulling your head out of your ass."

That's what you need when you're going to start a company or start a huge new project—a coach. A mentor. A source of wisdom and aid. Someone who can recognize a brewing problem and warn you about it before it happens. And someone who will quietly inform you that it's dark right now because your head is jammed up your own ass, and who will give you a few tips to quickly remove it.

You can make do without a cofounder. You can survive for a while without a team. But you can't make it without a mentor.

Find at least one person who you deeply trust and who believes in you. Not a life coach or an executive leadership consultant, not an agency, not someone who's read a lot of case studies and is ready to charge you by the hour. And not your parents—they love you too much to be impartial. Find an operational, smart, useful mentor who has done it before, who likes you and wants to help.

You will need to lean on them when you start a company. Or even when you launch a project within a big company.

Just don't think the latter option will be easier, that you'll be able to skirt the difficulties of a startup by starting it within someone else's corporation. Big companies are not a shortcut. Their spacious, attractive offices are littered with the skeletons of innovative little projects that died because they were set up for failure from the start.

You should only create a "startup" within a big company if they can offer you something unique—some technology, some resources that you can't get access to anywhere else. And you have to make sure that you have the right incentives, org structure, and management air cover in place to give you a fighting chance of success.

You have to remember that you'll be the proverbial gnat on the elephant's ass, in competition with other vastly larger revenue streams, trying to earn a place at the table. Even if you're at a billion-dollar company with nearly infinite resources, you can't expect those resources to be funneled your way without a fight. And you can't expect people at the company to take a risk on your project—to join your team and leave another, more established and respected area of the business—without some real reward. Same goes when you're recruiting outside the business—when you're trying to get someone to join your small new project at a big company versus going to a startup, you'll need a really good explanation for why it'll work and be worth their while. The calculus of risk and reward will have to make sense.

One reason we managed to put together an outstanding team to create the iPod was that our team could get relatively outsized stock and bonus plans that they couldn't get anywhere else at Apple. The other important reason was that we had Steve Jobs fully behind us. Those two things allowed us to recruit amazing people—even though we couldn't tell them what they'd be working on before they signed on—and survive the internal antibodies. Steve gave our tiny team an unfair advantage—gave us air cover and dropped bombs if anyone messed with us. There were times when the internal antibodies at Apple tried to expel us from the organization—we'd constantly hear "We have other priorities, we'll help you if we have time." Or "Why are we doing this project—it's not core to our business." But as long as our team was making reasonable (or unreasonable but important) requests, the teams who were stalling us would get a call from Steve. "If they're asking for something, then give it to them for Christ's sake! This is very important for the company!"

Nobody wanted to get that call. They learned not to throw themselves in front of a speeding train.

So if you don't have a CEO who will go to bat for you, if you don't have compensation packages that will attract a great team, if you don't have the resources of a giant company but all of the overhead, then don't try to start your project inside someone else's business.

Your best option is probably to go it alone. Either let your idea die or start a real startup.

Many startups are founded by entrepreneurs who just left big companies. They saw a need, pitched their bosses, got rejected, then struck out on their own. I watched it happen with Pierre Omidyar at General Magic. In his spare time, he wrote up some code to let people auction off collectibles to each other. When it started picking up steam, he asked if General Magic would be interested in it. No, thank you, was the answer; that's a ridiculous idea. So he got a waiver from General Magic that said they claimed no rights to his work, quit, and started a small startup called eBay.

There were a lot of reasons for Pierre's success—perfect timing, a great idea, the will to follow it, the skill to implement it, the ability to lead. He also had one huge advantage that many people don't think about: he came from a startup. He knew how it worked, had plenty of examples of what to do—and what not to do.

I've seen way too many people come out of the corporate world, decide to start a company, and be completely unprepared for what it takes. If they've never been on a small team starting from scratch, they're often a fish out of water. They spend too much money too fast. Hire too many people. Don't put in the time, don't have the startup mentality, can't make hard decisions, are buried by consensus thinking. They end up making mediocre products or nothing at all.

Don't let that be your story. If you want to start a company, if you want to start anything, to create something new, then you need to be ready to push for greatness. And greatness doesn't come from nothing. You have to prepare. You have to know where you're headed and remember where you came from. You have to make hard decisions and be the mission-driven "asshole." [See also: Chapter 2.3: Assholes: Mission-driven "assholes."]

So do the work. Know what you're getting into. Trust your gut.

And when the time comes, you'll be ready.

Chapter

4.3

MARRYING FOR MONEY

Every time you raise capital, you should think of it as a marriage: a long-term commitment between two individuals based on trust, mutual respect, and shared goals. Even if you take money from an enormous venture capital (VC) firm, everything ultimately comes down to the relationship you form with a single partner at that firm and whether your expectations are aligned.

As in marriage, you can't just throw yourself at anyone who shows a little interest. You have to take time to find someone you're compatible with—who doesn't play games or pressure you too much—and make sure it's the right moment for you to settle down. You don't want to get married when your company is so young that you don't know who you really are or what you want to become, or just because all your friends are doing it, or because you're scared that if you don't make a commitment now you won't be able to find another relationship.

You also have to understand your partner and their priorities. For example, a VC is beholden to the limited partners (LPs are large-scale investors or entities like banks or teachers' unions or very wealthy families) who fund that VC—so they may push you to sell or go public before you're ready in order to show value to the LPs. And a company with a VC arm, like Intel or Samsung, may use their investment in your company to wrangle a better business deal for themselves at your expense. Even if your funders have your best interests at heart—even if your mom is your angel investor—that doesn't necessarily mean her money comes risk-free or with no strings attached.

••

The reason venture capital exists is to facilitate transactions—you need money, they give you money. But the reason it works is relationships—the back-and-forth between you and a VC during the pitching process, the way a VC helps you recruit execs or run your board after the deal, the connections they offer for your next round. Venture capital is not fueled by money. It's fueled by humans.

And the rules for every successful human relationship are the same: before you can jump headfirst into a major life-changing commitment, you need to get to know each other. Trust each other. Understand each other.

That means you have to be ready to be scrutinized, to be examined, and—most likely—to be found wanting. You might hear "no" a dozen times before you find "the one." It's like a particularly brutal form of dating—but instead of asking to buy them a drink, you're begging them for money. It ain't fun.

Another thing: you'll never hear "it's not you, it's me." It's always you. It's your company, your ideas, your personality that will be judged.

It's hard to be exposed like that; it's hard to open yourself up. And that's true even if everything's gone crazy—even if it seems like anyone with half a pitch deck is getting funded.

Like how it was in 1999. Or how it is right now in 2022.

The world of investment is cyclical. The funding environment is always shifting back and forth from a founder-friendly environment to an investor-friendly environment. It's like the housing market—sometimes it's good for sellers, sometimes for buyers. In a founder-friendly environment, there's so much money flowing into the marketplace that investors will fund just about anything because they don't want to miss out on any deals. In an investor-friendly market, there's a lot less capital to go around, investors are pickier, and founders get worse terms.

And then sometimes there's a crazy market when it feels like cash is raining from the sky, all the rules have been thrown out, and it's never going to end.

But it will end. Just like it ended in 2000. There's always a reversion to the mean. And even when it's crazy, it still won't be easy. You'll still have to work for it. Details will still matter. Even if it looks easy, it never is. There are just varying degrees of difficulty—from damn hard to nearly impossible.

So before you start this process, you first have to know yourself and be sure of what you're asking for. Because you won't get a second chance at your first round. You have to be serious. You have to prepare. And you have to know what you're getting into.

The first question you should ask yourself is the most basic one: does your business actually need outside money right now? For many early pre-seed-stage startups, the answer is "no" surprisingly often. If you're still researching, still testing things out, making sure your idea has legs, then you don't need to immediately leap to financing. Take your time. Get comfortable with delayed intuition.

If you do think you're ready to take money, then what exactly do you plan to use that money for? Do you need to build a prototype? Recruit a team? Research an idea? Get a patent? Petition local government? Fuel a partnership? Create a marketing campaign? What's the minimum amount necessary to meet your needs now, and how much will you require later as those needs change?

Once you understand that, you can think through whether you have a business that investors will want to invest in. It's not a given that your company is right for venture capital. Most big VCs are surprisingly risk averse—they won't invest in startups that can't prove they're already on a clear growth trajectory. VCs have been trained by the internet age to expect numbers before they invest: growth rates, sign-up rates, click-through rates, unsubscribe rates, run rates, all the rates. And VCs have bosses to report to—their LPs, the people and organizations who give *them* money. They need to show that they're making wise, highly profitable investments with the right management teams.

If they do invest, many large VCs will assume you need a huge infusion of cash right away so you can quickly show a big return. Those expectations and that timeline don't make sense for many startups.

So don't assume you have to immediately start chasing brand names. You have a lot of options—giant VCs who invest in hundreds of companies and give out tens and hundreds of millions of dollars; smaller niche or regional VCs who invest in a handful of businesses; angel investors who can make a small contribution to get you started and ready for bigger VCs later; and companies with investing arms that are looking to use your product or wrangle a business deal. All these options exist all over the United States and all over the world, not just Silicon Valley. There's money everywhere now.

But regardless of what source of capital you choose, everything ultimately comes down to the individuals you'll work with. Even if you get a meeting with the biggest firm in Palo Alto, you won't be meeting with the whole firm. You have to impress and form a relationship with one person in that room: the partner. That's who will decide the terms of your agreement, who will be on your board. That's the person you're marrying.

I once worked with an entrepreneur who was pitching a huge brand-name VC. After a great meeting, the VC said they were in—they'd send over a term sheet right away. A week went by, then another. Then the managing partner started playing games, trying to reduce the valuation. He'd ignore the entrepreneur for a week, then come roaring back with more questions. And it just kept going—for four, five, six weeks.

In the meantime, the entrepreneur started talking to other VCs. And one of those firms sent over a term sheet the next day.

The entrepreneur had to make a tough call. Wait for the biggest player to get around to you or go with a lesser-known but much more enthusiastic investor? Who would be a better partner? Who would be more helpful in the long term?

So the entrepreneur called the big-name VC and broke the news: they'd signed with another firm. The partner was livid—started screaming the kind of stuff you hear from villains in eighties movies. "What the fuck! You don't do that to me!" He slammed the phone down and was never heard from again. And I mean that literally—this partner still won't talk to the entrepreneur. Pretends they don't exist. Avoids them at parties.

But being on the partner's shit list was much better than taking his money and having that asshole standing on their startup's neck. The entrepreneur had dodged a bullet. All the stalling had been a tactic to rock the entrepreneur's confidence, grind them into accepting worse terms. And when that stupid game backfired, one of the most well-known names in the Valley turned into a bitter child. Not someone you want to get into bed with, never mind marry.

Remember, once you take money from an investor, you're stuck with them. And the balance of power shifts. A VC can fire a founder, but a founder can't fire their VC. You can't divorce them for irreconcilable differences.

And if things go south, you can end up in an estranged marriage—still legally tied together, but never speaking. When a VC writes off your company, they basically ignore you. Won't help you. Won't connect you to other VCs. Won't speak up for you to partners. They'll stand on the sidelines as your company goes bankrupt.

So you should always pay very close attention to how a VC treats you when they should be on their best behavior—when you're getting along in the process and it seems like you might come to an agreement. If they start screwing around with you then, it should always ring alarm bells in your head. Here are a few other warning signs:

- VCs who promise everything under the sun to get you to sign, then don't deliver. Often they'll repeat themselves over and over, telling you just how much personalized attention you'll get, how much help, how much this or that. Make sure to talk to other startups who've worked with them to find out what they actually offer when they're not in sales-pitch mode.
- VCs who force the timing—who give you a term sheet to sign right here, right now—to make you feel panicked. A VC once gave me a term sheet as I was walking out of the meeting and pushed me to sign it on the spot. I asked if this was a used car dealership and told him I'd only sign after I read the terms.
- Greedy VCs who will invest only if they can take an outsized piece of your company. Typically a VC needs between 18 and 22 percent to make their model work—step carefully if they begin asking for more. And don't assume they're the only game in town—if your gut is telling you to keep looking, then keep looking.
- Some VCs court very inexperienced startups with the intention of pushing them around and telling them what to do rather than allowing the founder and CEO to run the company. Mentorship and advice is one thing; orders that must be obeyed are another.
- Sometimes a potential investor sees something interesting in your company—maybe you haven't received money from the right VCs, or you're tight on cash, or you have an incredible breakout success. So they'll propose a really sweet deal, but their terms will screw the other investors who have gotten you this far. We see lots

of big and small VCs trying to gain an edge by not playing fair—they may overdilute previous investors or put terms in your deal that scare off new ones. And if things aren't going well a couple of years down the road, they'll have no qualms about screwing you, too. So be careful when the terms aren't standard or seem too good to be true—sometimes it may appear that you're only giving away a small thing, but if it doesn't feel right in your gut, it may be that they're trying to get a foot in the door to start turning the tables against everyone else. They want to control your company sooner or later.

One thing that many founders worry about but which isn't usually a warning sign is if a VC has fired CEOs or founders in the past. Do your research—look at their track record. There are some well-known firms who are so focused on the company that they cut off founders' heads without giving them a second chance, but most VCs are generally hesitant to remove founders. Sometimes too hesitant. And those who do so infrequently will have a very good reason.

In any case, it's hard to generalize across an entire firm. It usually comes down to individuals. Just like everything else.

So when you reach out to pitch an investor, make sure you're reaching out to the right person. Talk to founders who have worked with the VC in the past—who've gone through tough times together—and find out which partner is operational and helpful and smart and which only cares about the money.

Try to get a warm introduction—either through another founder or your mentor or a friend of a friend. Even if it's a tenuous connection, it's better than nothing. The hardest way to get a meeting with a VC is by cold-calling them. And before you make the call, try to drum up a little press, get some good PR, so when the VC looks you up there's something to see.

Always remember that investors are swimming in pitches. Especially the big VCs, but also the smaller ones. You need some way to rise to the top and get their attention.

The best way to do that is with a compelling story. And knowing your audience. Even in Silicon Valley, most VCs won't be technical. So don't focus on the technology, focus on the "why." [See also: Chapter 3.2: Why Storytelling.]

It won't be easy to fit everything you want to say into fifteen slides—to make it flow in one smooth narrative, to make it compelling emotionally and rationally, to keep it high-level enough so that people can easily grasp the important points but not so high-level that it seems like you haven't dug into the details. It's an art.

As with all art, it takes practice. You'll probably suck at it at first. Pitching is hard. You'll need to constantly tune the deck up, change it around, tweak and revise.

So you don't want your first pitch to be in front of the very top VC in your area. VCs talk to each other, so if one casts you aside, the others in that class may pass as well. If you can, pitch a "friendly" VC first—one that will give you feedback and help you improve and then, hopefully, welcome you back a second time.

Remember that you don't have to come in perfectly polished for that first meeting. You can say, "I'd like to give you an early look at this. Maybe it would be of interest to you. I'd love to get your comments on it." Listen to their feedback and learn from it. You don't have to take every word of advice or criticism, but you should understand the reasons behind it and adjust accordingly.

Once you understand the pieces on the board, you can better plan your move. You can tailor the story to the individuals you'll meet. You'll start to feel ready.

Just don't forget the other factor that can sneak up and kick your ass: time.

It will take longer than you think to get funding. Expect it to be a 3–5 month process. It may end up being faster than that—especially in a founder-friendly environment—but I wouldn't gamble on it. Too many companies wait until they're about to run out of money, then hit an air gap and are near bankruptcy before desperately grabbing whatever funding they can get. Always start the pitching process when you don't actually need money. You want to be in a position of strength, not buckling under the pressure and making bad choices. You should also remember to watch out for the holidays—August, Chinese New Year, Thanksgiving through New Year. People forget that VCs take vacations, too.

Here are a few more tips to keep in mind as you're going through the process:

- Don't play games. Just as you don't want an investor who plays games with you, they will lose interest in you if you're not up front and honest with them.
- Listen to people's feedback on your pitch and your plan, change it when it makes sense, but hold on to your vision and your "why" and don't rearrange yourself based on the whims of every investor you talk to.
- Be clear with investors on how much money you need and exactly how you're going to spend it. Your job will be to create value for investors and ensure you're hitting major milestones to raise the valuation of the business. That way the next time you raise money you won't dilute existing investors, employees, or yourself.
- Entrepreneurs think their valuation should always increase, even when they don't hit the milestones they set out for themselves. But investors are running a business—if you don't deliver, you don't get a valuation increase and your own stock position dwindles. You may also need to dilute more because you'll need to give extra stock to employees to retain them through the hard times.
- Don't assume you'll get a valuation like other companies around you. Every investment stands on its own.
- Investors don't like to see when founders or executives are "fully

vested"—they want to make sure you have skin in the game. You may have to "re-vest" some of the shares you have already to demonstrate your commitment to new investors.

- Remember that investors are going to want references—they'll want to talk to your customers as part of their due-diligence process. So have a data room with a collection of files that will make it easy for them.
- In subsequent meetings, be up front with your risks and ways to mitigate them, who you need to hire, and the major challenges ahead.
- Try to get two similarly influential investors to balance each other out. All VCs know each other, all VCs talk—and nobody wants to piss off their potential partners. So if one of your investors starts playing games, the other can tell them to cut the crap. Your business may not matter all that much to them in the long run, but typically nobody wants to ruin their reputation among other VCs and especially the LP community.

Finally, remember that even if you have an incredible meeting—everyone loves the pitch, you love the investors, the room is practically vibrating with good energy—even then the people you met with will have to go back and convince the investment committee to give you money.

For every VC that process is different, so keep asking the question: What's the next step to get us to a yes? What's the next step? What's the next step?

It's like playing chess. You always have to think two moves—and two investment rounds—ahead.

Even if you're not interested in VCs yet. Even if you're just looking for an angel.

The best thing about angel investors is that they aren't beholden to LPs. They simply believe in you. They want to help you. And there's nobody looming over their head demanding an immediate profit.

Angel investors are typically much more willing to take a risk, so they may fund you earlier than a VC would and give you much more leeway and time to figure out your company without pushing nearly as hard.

This can be great. Or the lack of constraints can be a hammer to your knees. [See also: Chapter 3.5: Heartbeats and Handcuffs.] Or the guilt can take a hammer to your heart.

When I was twenty years old, I borrowed money from my uncle to start ASIC Enterprises, my startup that created processors for the Apple][. But then Apple stopped making Apple][s and ASIC disintegrated, along with my uncle's money. I felt awful—truly terrible—for years. But my uncle was very up front with me. He told me he knew he was making a bet, a bet on me, and that he was likely to lose it.

Fifty percent of marriages fail, but 80 percent of startups do.

If you start a company, the odds are against you. So you'll need to get over the mental anguish of failure and losing other people's money. If and when the time comes, you'll have to be honest and open about it; you'll have to admit what went wrong and how you learned from it.

But there's nothing you can say that will make it easier. A VC's money is one thing; your mom's is another. If you take money from family and friends, you'll have to work just as hard, if not harder, than you would if you took cash from a VC. And you'll have to reckon with the chance that you'll have to walk back to them empty-handed.

Even as I was starting Nest, I didn't want that burden. I refused to take money from Xavier Niel, a good friend and incredible entrepreneur who founded Free, a French internet service provider. I wasn't a twenty-year-old kid anymore and Xavier was in a very different spot financially than my uncle. But I didn't want Xavier to think I was after his money. And I remembered that feeling of failure, of telling someone I cared about that their money was gone. So Xavier kept asking, and I kept saying no.

Finally we were onstage together just after Nest launched—in front of ten thousand people—and he turned to them and said, "He won't let me invest!" At that point Nest was doing well and wasn't nearly as risky, so I agreed to finally take his money. It turned out great in the end, but I didn't want anything to poison our relationship in the beginning. Nest was stressful enough.

No matter which route you take—VC or angel or strategic or bootstrap—starting a company is hard. Getting money is hard. There are no shortcuts, no easy path, no room for dumb luck.

But if you do it right, if you choose the right people, then you'll genuinely like your investors and they'll help you through the tough times that always come with a startup. They'll be there in sickness and in health and you'll end up in a happy marriage. Maybe even a few of them.

After that, all that's left to do is build a business.

Chapter

4.4

YOU CAN ONLY HAVE ONE CUSTOMER

Regardless of whether your company is business-to-business (B2B) or business-to-consumer (B2C) or business-to-business-to-consumer (B2B2C) or consumer-to-business-to-consumer (C2B2C) or some-yet-unimagined acronym, you can only serve one master. You can only have one customer. The bulk of your focus and the whole of your branding should be for consumers or business—not both.

Understanding your customer—their demographics and psychographics, their wants and needs and pain points—is the foundation of your company. Your product, team, culture, sales, marketing, support, pricing—everything is shaped by that understanding.

For the vast majority of businesses, losing sight of the main customer you're building for is the beginning of the end.

••

Before the days of Linux servers, when Windows servers dominated the landscape, Apple decided to try B2B. To build their own servers. The project started just before I joined—Apple was desperately trying to crack the code on how to increase computer sales and bring in more developers. Corporate users needed to run all kinds of enterprise software from servers, so the quintessential consumer brand built a server for the enterprise.

It was a failure. It wasn't that the technology was too difficult—that was actually the easiest part. B2B just wasn't in Apple's DNA. They didn't have the marketing, sales, support, or developers. And corporate Chief Information Officers (CIOs) were accustomed to the countless enterprise-level services that Microsoft and Windows offered. Apple's hardware was a tiny piece of the puzzle that these CIOs needed to make a purchase decision. The server team twisted themselves into knots trying to force an unnatural pairing—an Apple tree trying to grow an orange—until the iPod took off and saved the company and the server project was blessedly cancelled.

Steve Jobs was clear about the lesson he'd learned and made sure we all learned it, too: any company that tries to do both B2B and B2C will fail.

Is your customer Jim the Millennial, who saw your ad on Instagram then bought your product as a Christmas present for his sister? Or is it Jane the CIO of a Fortune 500 company who answered a cold email from your sales team, negotiated pricing and different product features for months, and now needs a team of customer success agents to train the five thousand employees under her? You cannot hold both those people in your head at the same time. You cannot make a single product for two completely opposite customers—for two different customer journeys.

Not when you're making technology. Or services. Or a store. Not even if you're making dinner.

It's a hard rule.

But for every rule there are exceptions. Just because you start off as B2C doesn't mean you can never work with the enterprise in any way. And a small number of very specific companies can split themselves down the middle and do just fine: travel businesses like hotels and airlines, retailers like Costco and Home Depot (their big innovation was taking a B2B product and opening it up to B2C). Financial products and banks can be both B2B and B2C, since some households run like a small business.

But even these companies have a fully B2C brand. That's the other rule: if you cater to both, your marketing still has to be B2C. You can never convince a regular person to use a B2B product that's obviously not meant for them, but you can convince a company to use your product if you appeal to the human beings inside that company.

That's ultimately how, despite itself, Apple found itself firmly in the enterprise.

After the iPhone launched, CIOs were slow to clear it for business use. And while CEOs usually cede anything IT-related to the CIO, this time the CEOs rose up and demanded change. They loved their iPhones. So did their employees. And they wanted to use them at the office.

It was Apple's success at creating something for consumers that led to their success in the enterprise. People fell in love with their phones and then wondered why the rest of their life wasn't that easy. Nobody wanted to deal with garbage corporate tools that required days or weeks of training to use. They wanted an easily understandable interface, fast speeds, and slick hardware.

One of the main pulls to create the App Store was actually from corporations. As businesses began adopting the iPhone, they reached out to Apple, asking to make applications for their employees and sales. If Apple wanted people to continue using their phones for work, they'd have to give the enterprise the ability to create its own apps. And so the App Store was born.

Now Apple has separate teams to handle all of their B2B business, but the products are never defined to appease B2B customers. By keeping itself pure as a B2C company, Apple was able to tack on B2B without significantly altering its priorities or its marketing, or knocking its core business out of whack.

After Steve set the rules, Apple followed them. They know how the game is played.

But what happens when the game changes? What if it's not just

B2B and B2C anymore? What if there are new marketplaces, new services, new business models, new acronyms?

One of the companies I work with is DICE. It's a next-generation music discovery and ticketing platform that's B2B2C. And in its early years, DICE was torn in three directions by its three customers: music fans (consumer), music venues (business), and musicians/managers (business). On the one hand, DICE made most of its money from venues, so maybe its tools should cater to them. On the other hand, it wanted to create a great experience for fans. On the other hand—none of this would be possible without the artists, so maybe they should be the focus.

DICE needed to attract all three customers. It needed to keep all three happy to be successful. But DICE only had one team and one product. And every time it made concessions to the venues, the experience for the fans and artists suffered. When it tried to please the artists, the venues complained.

My advice was simple: Nothing has changed. The rules still apply. You gotta pick one. And the whole reason you started this company was to get rid of scalpers, to make an amazing experience for fans. You're B2B2C, but don't lose your mission as you navigate your acronym. The Bs matter, but without the C you have nothing.

Now it's their "Golden Rule": Our Only Customer Is The Fan.

And they make sure the venues and artists believe it, too. They remind them, continually, that if DICE does right by the fans, everything else will follow. The artists, the venues, DICE—they all have one master in the end: the person buying the concert ticket. The human being who just wants to see a great show.

That's the thing to remember about B2B2C—it doesn't matter how many businesses are involved: ultimately it's the end consumer who carries the business model on their back.

But companies forget. It happens most often when a company evolves from B2C into B2B2C. They usually start off with no business model, no way to make money, just a lot of customers using their product for free. But free is never really free. Eventually many of these companies realize their most lucrative option is to sell users' data to big business. That means bolting on B2B sales so they can resell customer data hundreds or even thousands of times. That's the story of Facebook and Twitter and Google and Instagram and many, many others.

But that story can end badly. When attention and focus shift away from the consumer and toward the businesses bringing in the real money, companies go down some very dark alleys.

And it's always the consumers who suffer.

So do not lose your focus. Do not think you can serve two masters. No matter what you're building, you can never forget who you're building it for. You can only have one customer. Choose wisely.

Chapter 4.5

KILLING YOURSELF FOR WORK

There are two kinds of work/life balance:

1. **True work/life balance:** A magical, quasi-mythical state where you have time for everything: work, family, hobbies, seeing friends, exercising, vacationing. Work is just one part of your life that doesn't intrude on any other part. This kind of balance is impossible when you're starting a company, leading a team that's trying to create innovative products or services on a competitive timeline, or just experiencing crunch time at work.

2. **Personal balance when you're working:** Knowing you're going to be working or thinking about work most of the time and creating space to give your brain and body a break. To reach some level of personal balance, you need to design your schedule so you have time to eat well (hopefully with family and friends), exercise or meditate, sleep, and briefly think about something other than the current crisis at the office.

To withstand a complete lack of true work/life balance requires a clear organizational strategy. You need to prioritize. It's important to have everything you need to think about written down and have a plan for when and how you'll bring it up with your team. Otherwise it will swirl around in your brain endlessly, killing any meager chance you have of relaxing your shoulders for a minute.

••

Here's my advice: do not vacation like Steve Jobs.

Steve would typically take two weeks off, twice a year. We'd always dread those vacations at Apple. The first forty-eight hours were quiet. After that it would be a storm of nonstop calls.

He wasn't tied up in meetings, worrying about the day-to-day, so he was free. Free to dream about the future of Apple at all hours of the day and night. Free to call and get our thoughts on whatever crazy idea just occurred to him—what about video glasses to watch movies from the video iPod? Yes? No? He'd want us to give our perspective right then or find answers fast so he could refine his thinking.

He worked harder on vacation than he did in the office.

That kind of crazy, nonstop focus sounds like just another Apple legend. The kind of thing only a mad genius would do. But it's not, really.

Steve took it to an extreme, but a lot of people can't get work out of their heads. I can't. I'd hazard to say that most people can't, especially when there's a lot on the line. It's not just CEOs and executives—everybody has crunch times. There's just too much to do and you know more is coming, so even when you're not doing the work, you think about it.

And sometimes that's okay. Really. Sometimes it's your only option. But there's a world of difference between racking your brain, ruminating all night about a work crisis, versus letting yourself think about work in an unstructured, creative way. The latter gives your brain the freedom to stop hammering away at the same problems with the same worn-down tools. Instead, you let your mind rummage around to find new ones.

Sometimes I thought that's why Steve took vacations—not to relax, not to avoid Apple, but to give himself a nice long time to rummage as he spent time with his family. Instead of trying to find true balance or allowing anyone else to find it, Steve ran full tilt. He let Apple become all-consuming in a way that pushed everything else in his life, except his family, to the periphery.

Most people have experienced that kind of complete collapse of work/life balance in critical moments when the pressure's really on. But it's how Steve lived. And if you're not Steve Jobs—if you have to think about work all the time but you don't *want* to think about work all the time—then you need to have a system.

You have to find a way to stay sane—to manage the inevitable swirling morass of tasks and meetings and plans and questions and problems and progress and fears. And you have to architect your schedule so your body and brain don't get burned out or bloated beyond recognition. I say this as someone who's been there—I physically and mentally fell apart at General Magic. Humans cannot survive on stress and Diet Coke alone.

But General Magic was one thing—that was early in my career, when I got caught in the explosion but didn't cause it. Apple was something else entirely. It's hard to describe the pressure of my first few years there. Especially at the very beginning, when I was running my startup while doing Apple contract work, trying to maneuver that into a way to save my team at Fuse. The stress only got cranked up after I began working on the iPod full-time.

In the beginning, it was a side project at Apple. But in the months and years that followed, the iPod became as important as the Mac, sometimes more important. The company was holding its breath, waiting to see if we'd succeed. Not only did we need to build this completely new thing, but we needed to build it incredibly quickly, up to Steve Jobs's exacting specifications, make it beautiful and delightful in a way that would remind everyone of what Apple could be, and then have it be a roaring commercial success.

After getting Steve's green light, I walked into the building in April 2001 knowing that we were going to have to design and build the iPod by the next holiday season—seven months away. And not because Steve set a crazy deadline. I did. Steve figured it would take 12–16 months. Everyone did.

Nobody believed we'd be able to pull it together in time to get it to customers by Christmas. But I had just come off four years at Philips, where more than 90 percent of projects got cancelled and killed. If you didn't make your mark fast enough or your project ran into issues or dragged on, Philips corporate would descend upon you, ready to "save the business" from your mistake or steal it out from under you. [See also: Chapter 2.3: Assholes.] I didn't know if it would be the same at Apple or not, but I couldn't take the risk.

Just like I couldn't risk Sony launching a music player that Christmas and eclipsing us or risk getting caught in internal Apple politics. We were a tiny little team sucking resources away from the core business, which was under huge financial pressure to succeed. Other groups did not like that. They did not like us. I could feel their eyes on us and their daggers out.

So we had to prove ourselves. We worked relentlessly. My job was building and leading the team to create the iPod from scratch, including building the team itself. I had to be close to the day-to-day design and engineering work but also manage the expectations of the executives but also work with sales and marketing to make sure not to repeat the mistakes of Philips but also go to Taiwan to check on manufacturing but also make sure my team was dealing with the stress but also debate with Steve and other execs daily but also occasionally attempt to sleep.

Keeping everything in my head was impossible. There was always a new crisis, a new worry that would pop up, supplanting whatever I was worrying about a second ago. There were just too many tiny moving parts, too many gears that needed to turn other gears that needed to turn other gears, a half-made clock constantly cuckooing in my ear.

I needed to calm down. I needed to find space. I needed to prioritize.

Everyone thought I was crazy—and many still do—but here's what I did: I took several sheets of paper with me everywhere. They had all the top milestones in front of us for each of the disciplines—engineering, HR, finance, legal, marketing, facilities, etc.—and everything we needed to do to reach those milestones.

Every top-level question that I had was on those papers. So when I was in a meeting or talking to someone, I could quickly scan it. What are my top issues? What issues do our customers have? What's the current roadblock for this person's team? What are the next major milestones? What date commitments did our teams make?

And then there was the best part—the ideas. Whenever someone had a great idea that we had to table for the moment—an improvement to the product or the organization—I'd write it down. So right next to the list of that week's to-dos and tasks, there was a working library of all the things we couldn't wait to begin. I'd regularly read them to myself and see if they still applied. It kept me inspired and excited and focused on the future. And it was great for the team. They saw that I paid attention to their ideas and made sure we kept thinking about them.

The only way for me to capture all of it—good ideas, priorities, roadblocks, the dates that people promised to deliver, and the major internal and external heartbeats ahead—was to take notes in every meeting. Longhand. Not on a computer. [See also: Figure 3.5.1, in Chapter 3.5.]

Writing by hand was important for me. I wasn't staring at a screen, getting distracted by my email. A computer or a smartphone between you and the team is a huge barrier to focus and sends a clear message to everyone in the meeting: whatever I'm looking at on my screen is more important than you.

Even just taking notes on a computer was a nonstarter. Sometimes when I'm typing, I just . . . type. Whatever I'm writing down doesn't make it all the way into my brain. But there was too much on the line for me to zone out, to not hear every word my team was saying.

The act of using a pen, then retyping and editing later, forced me to process information differently.

Every Sunday evening, I would go through my notes, reassess and reprioritize all my tasks, rifle through the good ideas, then update those papers on a computer and print out a new version for the week. Continually reprioritizing allowed me to zoom out and see what could be combined or eliminated. It let me spot moments when we were trying to do too much.

Those were the evenings when I'd realize why we were so overwhelmed—we had said "yes" to too many things and we needed to start saying "no." And then came the hard work of figuring out what had to be delegated, what had to be delayed, and what had to be crossed off the list. I was forced to prioritize based on what really mattered, as opposed to what was just top of mind. That let me keep my eye on the bigger goals and milestones ahead of us, not just the fires at our feet or whatever feature we were most excited about that day.

Then Sunday night I'd email the whole list out to my management team. Each item had a name attached to it. Everyone could look at the top of the list to see what I'd be focused on that week, what they were accountable for, and what the next major milestones were.

And every Monday, we'd have a meeting about it.

Everyone hated it. I literally watched people flinch when I'd bring out the papers, scanning them for the thing I'd been asking about for weeks. That thing I'd refuse to forget about because it hadn't gotten crossed off the list yet. *On June 3 you said it would be ready by the end of the month. It's now July—what's the status of this project?*

It wasn't micromanagement. It was holding people accountable. It was holding everything in my head at the same time. It was holding on for dear life amid the flood of everything I needed to remember.

It started as a one-sheeter. Eventually it grew to eight pages, ten pages. It was labor intensive. Arcane. Never-ending. But it worked. And eventually my team grew to appreciate it. It kept me (relatively) calm. It helped me focus. And nobody ever had to wonder where my head was at. Everyone always knew what mattered to me—they had my priorities in writing, updated, every week.

A lot of them have picked up the practice themselves, and the people who worked for them have done the same. Everyone dreaded the list, the email, the meeting—until they too had too much in their brains. Until they needed a way to manage it.

I'm not saying this will work for everyone. Far from it. Everyone needs to find their own system. But you do need to prioritize your tasks, manage and organize your thoughts, and create a predictable schedule for your team to access those thoughts.

And then you need to take a break.

A real break. Take a walk or read a book or play with your kid or lift some weights or listen to music or just lie on the ground, staring at the ceiling. Whatever you need to do to stop your mind from frantically spinning in circles about work. Once you have a way to prioritize your tasks, you need to prioritize your physical and mental well-being. And I realize that's easier said than done. Your startup or the project you're leading is your baby. And babies roll down stairs, eat extension cords. They need constant attention.

That's what work can feel like. Even if you take a vacation—and if you're starting a major project you won't go on vacation for a good long while—it's like leaving your kid with a babysitter for the first time. You're pretty sure they'll be fine, but, you know, you'll check just in case. And again in an hour. And maybe on the way home. Did you tell the sitter the baby sneezes when she's sleepy? Better call again.

Eventually, you get to trust the babysitter. You'll know your team can handle things without you. After a few generations of iPod were done, I took some real vacations.

I'd like to say I wasn't like Steve Jobs—that I focused on my family and my personal joys and took time to relax. But I didn't, really. I also spent all my time thinking about the future of the company—but in a different, less focused way than I did every day at the office. I rummaged.

I just didn't call or email anyone about it. We talked only if there was a real emergency.

Every time I left, I'd hand the reins over to a different person who reported to me. It's your problem now, buddy! It was a time for the team to step up and learn to do what I did. Vacations are a great way to build a team's future capabilities and see who might step into your shoes in the years to come. Everyone thinks they can do your job better—until they actually have to do it and deliver. So even if you're in a high-stress job, you need to take vacations. They're important for your team.

And they're a great time for you to try to sleep. For a long, long time, getting enough sleep for several days in a row was almost impossible.

I slept well, really well, before 1992. Before international email was a thing, let alone the internet and Twitter. Since then there has literally always been someone in some time zone somewhere wanting to talk to me at 4 a.m.

There will be no break unless you force yourself to take one. So do all the stuff they tell you to do before bed: no caffeine, no sugar, keep it cold, keep it dark, and for the love of all that's holy, keep your phone away from your bed. You're an addict. We all are. So don't make it too easy for yourself—charge your phone in another room. Don't be the alcoholic with a whiskey bottle in their nightstand (I wish I could say I do this every day, but hey—I'm human, too).

Then create time to breathe in your schedule. It's all too easy to go from meeting to meeting to meeting, all day, with no chance to eat or go to the bathroom, never mind time to rest. But you have to. And I mean that literally. You have to do it. Otherwise you will fall apart. We've all seen (or been) parents with newborns on the brink of total collapse—that's what it can feel like. Part of your job is not to go completely nuts at work and take it out on your team.

There's a reason Steve Jobs famously always walked between meetings, during meetings. It helped him think, stay creative, to rummage, but it also forced him to take some time to just . . . walk. To take a break from sitting in meetings, even if it was just for a few minutes.

So look at your calendar. Engineer it. Design it.

Lay out the next three to six months on paper.

Write down what a typical day looks like.

And what a typical week or two weeks look like.

Keep going for the next month.

And then the next six.

Now start to reengineer your day, your week, and your monthly schedule with time dedicated to feeling human. It could be ten minutes after lunch where you read an interesting Medium article or six months from now when you take a week under a palm tree. But you have to engineer your schedule to include these breaks and then hold the line when people try to schedule over them.

So what will you do every few days or every week or two?

Every 8–12 weeks?

Every 6–12 months?

In the long term, you need to plan some vacations. In the short term, here's what I recommend:

- 2–3 times a week—Block out parts of your schedule during your workday so you have time to think and reflect. Meditate. Read the news on some subject you don't work on. Whatever. It can even be tangential to your work, but it should not be actual work. Give your brain a second to catch up. Learn, stay curious, don't just react to the never-ending stream of fires to put out or meetings to attend.

- 4–6 times a week—Exercise. Get up. Go biking or running or weight lifting or cross-training or just take a walk. I started getting into yoga at Philips and I've kept it up for more than twenty-five years—it's been hugely helpful. You have to quiet everything around you and focus to do yoga poses properly. You become conscious of your body so you instantly know if you're off. Find something like that—where you'll notice if you're physically or emotionally at a breaking point and will have an opportunity to right yourself before it gets too bad.
- Eat well—You are an extreme athlete, but your sport is work. So fuel yourself. Don't eat too much, don't eat too late, cut down on refined sugars, smoking, alcohol. Just try to keep yourself from physically feeling like garbage.

And if all this seems great in theory but completely impossible in practice because you can barely keep up with your emails, let alone make time to go to the gym or block out whole months of your life, then you may need to add something else to your to-do list: an assistant.

If you're fairly high level (director or above) at a fairly large company managing a fairly large team, then you should consider an assistant. If you're a CEO of any company, you should absolutely have one.

A lot of young leaders are uncomfortable hiring an assistant. I know I was. It felt like an admission of weakness, a sure sign of a hoity-toity exec who's completely lost touch. And you don't want to take advantage of someone—force an assistant to do "busy" work you really should be doing yourself. And anyway, you can't hire an assistant before you hire an engineering director or fill that open position in sales—you have other priorities.

But as a leader, you also have a job to do. And if you're spending a large portion of that job scheduling meetings and sorting through emails, or—worse—failing to do so, then there's a problem. Everyone has met, or been, that kind of leader. This is the person who falls behind and accidentally ignores emails for two weeks, then books three meetings on top of each other and shows up to none of them. They're so overwhelmed with scheduling their work that they can't get any work done. These are the people who make themselves and their teams look bad. Who make their company look bad.

Do not become one of those people.

If you're worried about the optics, just make the assistant a shared resource. A capable assistant can support three, four, or even five people. Or they can be a resource for your entire team—help them schedule their travel or figure out their expense reports or do some special project. They can help everyone.

Just remember, there's no perfect assistant who will be able to read your mind immediately. What you're looking for is someone who won't gossip about you or the company but who will be friendly with everyone on the team so they can funnel worrisome gossip your way. You want someone who learns fast, only needs to be told once, and who, over time, can anticipate your needs and fix problems before they ever hit your desk. It can take three to six months for them to get a handle on how to be most helpful, but then it truly feels like you have a new superpower. It's like you've gained another limb, or another six hours in the day.

This person isn't just an employee. They're a partner. So don't fall into the stupid movie trope of using them as a lackey. My incredible, do-it-all, smart, and kind assistant Vicky once worked for a person who decided one day that he needed—absolutely needed—organic cantaloupe right away, even though he was in the middle of nowhere, and sent her on an hours-long hunt to find it. That is not how you treat a precious time machine who unlocks days, weeks of your life.

But sometimes even having an amazing assistant won't be enough. Sometimes the pressure and stress and never-ending list and never-ending meetings will become too much. In those moments, get the hell out. Take a walk.

Sometimes when I just knew things were headed inexorably downhill, I left the office, rescheduled my meetings, and said, "Today is just one of those days—don't make it worse than it already is."

There are moments where you simply cannot function as a human, never mind a leader, and you need to recognize them and walk out the door. Don't make a bad decision because you're frustrated and overworked—get your head on straight and come in fresh the next day.

None of this is revolutionary. You probably learned it in elementary school: write down a list of what you need to do, take a deep breath and some quiet time if you're upset, eat your vegetables, exercise, sleep. But you'll forget. We all forget. So grab your calendar and make a plan. You'll be working all the time for a while. That's okay. It's not forever. But you've probably been beating at your problems with the same hammer for too long—it's time for your brain to rummage around and find a crowbar. Or a bulldozer. Give your mind some time to breathe.

Then put away your phone before bed. And maybe do some yoga.

Chapter

4.6

CRISIS

You will encounter a crisis eventually. Everyone does. If you don't, you're not doing anything important or pushing any boundaries. When you're creating something disruptive and new, you will at some point be blindsided by a complete disaster.

It may be an external crisis that you have no control over, or an internal screwup or just the kinds of growing pains that hit every company. [See also: Chapter 5.2: Breakpoints.] Either way, when the time comes, here's the basic playbook:

1. Keep your focus on how to fix the problem, not who to blame. That will come later and is far too distracting early on.

2. As a leader, you'll have to get into the weeds. Don't be worried about micromanagement—as the crisis unfolds your job is to tell people what to do and how to do it. However, very quickly after everyone has calmed down and gotten to work, let them do their jobs without you breathing down their necks.

3. Get advice. From mentors, investors, your board, or anyone else you know who's gone through something similar. Don't try to solve your problems alone.

4. Your job once people get over the initial shock will be constant communication. You need to talktalktalk (with your team, the rest of the company, the board, investors, and potentially press and cus-

tomers) and listenlistenlisten (hear what your team is worried about and the issues that are bubbling up, calm down panicked employees and stressed-out PR people). Don't worry about overcommunicating.

5. It doesn't matter if the crisis was caused by your mistake or your team or a fluke accident: accept responsibility for how it has affected customers and apologize.

••

One of the core features of the Nest Protect smoke and CO alarm was called Wave to Hush. The idea was that if you burned breakfast, you wouldn't need to frantically wave towels or brooms at your smoke alarm to shut it up—you could just stand under it and calmly wave your arm a couple of times.

Wave to Hush worked beautifully. Customers loved it. More importantly, Nest Protect was really helping people. It wasn't just solving the annoying false-alarm problem. We heard amazing stories of families who escaped fires and avoided carbon monoxide poisoning. We were immensely proud of this product and the lives and homes it saved.

And then, months after launch, during routine testing in our lab, a single flame grew much larger and higher than we'd ever seen. It rose, it danced . . . it waved. It fucking waved. And it hushed the alarm.

I'm not sure if I actually said "Okay, nobody panic," but I definitely thought it. My heart sank. It felt like I'd gotten punched in the gut. We had to whip out the crisis playbook: first understand what level of issue this is. Was it replicable? Was it a fluke of one messed-up test? Was it real? And if it was real, was it likely? One-in-a-thousand chance or one in a billion? Because if it was real—and it was dangerous—then the next steps could be brutal: a product recall, alerting customers, notifying the authorities. Or, infinitely worse, this crazy flame could appear in an actual house fire. It could shut down our alarm just when someone needed it most.

We had to work furiously to parallel-path every possibility:

1. We needed to recall every single Nest Protect. This might kill our product, our brand reputation, and all our sales.
2. We can solve this with a software update.
3. This is just a testing error.

This wasn't a moment to stand back and let the team figure out what to do on their own. I needed to make sure people knew exactly what they were working on and had the tools to find solutions as fast as possible. I had to command and control.

In a crisis, everyone has their job:

- If you're an individual contributor, you need to take your marching orders and start marching. Do your core job while continuing to look for and suggest other options to solve the issue. Try not to speculate or gossip. If you have concerns or suspicions, report them up the chain, then get back to work.
- If you're a manager, you need to relay information from leadership without overwhelming or distracting your team. Check in with the team a couple of times a day—try not to harass them more than that (hourly messages just freak everyone out). You need to be there for them, not just to ensure that the work is getting done, but also to make sure they're okay. You're the first line of defense against burnout. The pressure, stress, red-eyes, and bad food in the middle of the night will get to people. You may need to give everyone a break—even during a crisis.

  Remember to set expectations and limits. You'll probably have to work over the weekend. Okay. That happens. But tell your team what the plan is: we'll work hard on Saturday but everyone needs to get out of the office at 5 p.m. and then we'll have a check-in on Sunday night.

- If you're the leader of a broader group or company, you probably spent years of your life unlearning the tendencies of micromanagement. Well, if you're in a crisis then it's time to be a micromanager again.

You'll need to dig into the details—all the details. But you can't make every decision on your own or fix everything single-handedly. You have experts, so you'll need to delegate to them. Agree on the micro-steps that need to be taken, but allow them to take those steps without you. Schedule check-ins in the morning and at the end of the day and instead of getting the usual weekly or biweekly reports from your team, start going to their daily meetings. You have to be in there, listening, asking questions, and getting necessary information in real time. You might have to be the conduit of that information to the rest of the company, to investors or reporters or whoever else is watching this situation like a hawk. You need to be able to answer *their* questions. You need to keep up their confidence that you're getting somewhere.

Clear your calendar of nonessential meetings. Focus entirely on fixing the problem. And don't let yourself get knocked off balance—you're human. Don't make things worse by losing your mind and ignoring the things you need to keep your head on straight. That might be exercising or resting or having dinner with your family or lying on the floor under your desk for ten minutes quietly singing show tunes. Whatever you need. And remember, your team is human, too—people need to go home. They need to sleep. They need to eat. And they need to feel like things are getting better.

So keep their focus on your solutions, not who's to blame for getting you into this mess in the first place. Everyone will be going through the hypotheticals—what if it was this team's fault? Were they cutting corners? Gossip will be flying, along with accusations. But getting to the bottom of the screwup is not your team's job. It's not even your job. Not in the beginning.

You'll get there eventually, but you need to dig yourself out of your hole first. You need to solve for what went wrong and what you're going to do about it, then circle back for the why.

Don't forget that even when everyone calms down after the initial shock wears off and gets back to work, they're probably still freaking out internally, just like you. Especially if the task to find a way out of this disaster now rests on their shoulders. Make sure people who are struggling have an open line to talk to you or their manager. Command and control doesn't mean decree and ignore.

You're landing a dozen jets on an aircraft carrier at the same time—while giving press briefings and occasional therapy sessions. You will be incredibly worried, but you can't tear your hair out—I strongly recommend being bald already. The only thing you can do is calmly say, "Yes. I'm worried. Just like you. This is scary. But we'll get through it. We've faced other challenges in the past together and succeeded. Here's the plan."

That's what I said over and over at Nest. It became like a mantra: We'll get through this. We've done it before. Here's the plan. We'll get through this. We've done it before. Here's the plan.

Thankfully we never saw that bizarrely tall, narrow flame in the field—only in testing. It turned out to be one of those fluke things that we couldn't have anticipated or designed around. Nobody's fault. And it was very, very unlikely to happen in the real world. But that didn't matter.

The solution was to pull Nest Protect off shelves while we investigated and turn off Wave to Hush with a software update. You could still silence the alarm from your phone, but you couldn't wave at it. And we told customers exactly what happened. No cover-up. Mea culpa and here's a refund if you want one.

And it worked. Nest Protect and our brand survived.

There is always a temptation to obfuscate or couch everything in legalese—to say "mistakes were made" but never admit they were yours. This will not work. People will figure it out. And they will be pissed.

If something is your fault, tell them what you did. Tell them what you've learned from it. And tell them how you'll prevent it from ever happening again. No evading, blaming, or making excuses. Just accept responsibility and be a grown-up.

Every failure is a learning experience. A complete meltdown is a PhD program.

You'll get through it. Just remember that you don't have to get through it alone. In moments of crisis, it's critical to talk to someone who can give you useful advice. No matter how much you know, how good you are, there is always a person out there who can help you unlock a potential solution. Someone who's done it before and who can show you the way out of the tunnel.

Sometimes the terrible, unsolvable, unpredictable crisis you're facing is actually something that most growing companies face and for which there's an obvious solution that you just can't see. You might simply be growing really fast and need to codify your culture and add a management layer and start sending out meeting notes differently. [See also: Chapter 5.2: Breakpoints.]

So whenever you see the water rising, talk to your mentor. Or your board. Or your investors.

It's your responsibility as a leader not to try to deal with a disaster on your own. Don't lock yourself in a room, alone, frantically trying to fix it. Don't hide. Don't disappear. Don't imagine that by working for a week straight and not sleeping you can solve the problem yourself and nobody ever has to know. Get advice. Take deep breaths. Make a plan.

Then put on your rain boots and walk into the tidal wave.

The silver lining is that once the crisis is past—assuming you survived it, of course—you'll have a team that's gone through hell and back and is stronger for it. You'll have time to go figure out the why—why did this happen in the first place? And what can we do so it doesn't happen again? That may mean someone gets fired or the team reorganizes or the way you communicate with each other drastically changes. The process may be lengthy and unpleasant.

But after it's over you should celebrate. You should have a party. And you should tell the story.

The most valuable thing you'll take out of any crisis is the tale of how you were almost swept away, but the team pulled together and saved the day. That story needs to enter into the DNA of your company so you can always return to it.

There will be more disasters in your future. There will be many moments when everything falls apart. But if you can keep telling that story, no crisis to come will ever feel quite as bleak as the very first one you conquered. Because you can always turn to your team and say, "Look. Look what we survived together. If we got through that, we can get through anything."

It's a useful corporate tool to remind people what can happen, what you've learned, and how to avoid similar disasters in the future. The story is handy managerially and as a cultural touchstone. But, most importantly, it's true: your team got through this. Now they can get through anything.

Part

V

BUILD YOUR TEAM

By the time I left Nest in 2016, the company stretched across three buildings in Palo Alto and two more in Europe. We had almost one thousand employees, multiple product lines, ever-expanding sales partnerships in multiple countries, millions of customers, giant posters espousing our company values on the walls, black-tie holiday parties. But even with the slings and arrows of acquisition and rapid growth, Nest still felt like Nest.

And that's because of exactly one thing: the people.

The source of all Nestiness—the key to our success—was the human beings we hired, the culture they created, the way they thought and organized and worked together. The team was everything.

Forming that team and shepherding it through its many transitions is always the hardest and most rewarding part of building anything. And with Nest that was true from the first moment—before we had customers, before we even had a product.

When all we had were the squirrels.

They'd wander into our meetings a lot. And of course the rain was also a problem—we'd often have to cover the floor in buckets. The garage door made a ridiculous racket whenever the wind blew, there was exactly one ghastly pink marble bathroom for the whole team, and the beaten-up eighties-era chairs were truly terrible, especially the big pleather executive ones. I don't think there was a chair in the entire place that had all four legs on the ground.

It was exactly what we wanted.

It was the summer of 2010 in Palo Alto and the garage we'd rented was bordered by the vast, beautiful campuses of tech giants and countless shiny startups luring in employees with promises of plush offices and free beer and flexible work hours.

But none of that mattered to us. Matt and I were serious, focused, and we were hiring people who had that same sense of purpose, who wouldn't be dazzled by glitz or glamour or in-office pool tables. [See also: Chapter 6.4: Fuck Massages.] We were having plenty of fun, but none of us were screwing around.

By that point the team was around ten to fifteen people, the very beginnings of Nest.

A lot of the early employees were from Apple. A few I'd known since General Magic. Another since college. Our VP of marketing was a friend of a friend from Philips. Most of the team already had a ton of success in their careers.

But we all balanced precariously on the same crappy chairs. Furniture and snacks and decor all take money and, more critically, time. Someone has to sit and think about whether we should get a brown couch or a blue one, which fruit, which cheese, which beer. And we were not going to waste a cent or a minute on anything that wasn't critical to the business. We were going to show our investors that this was a world-class team who could perform miracles on a shoestring. With every squirrel invasion and ceiling leak our team was declaring that we were the opposite of every startup in the Valley that was lavishing money on their offices and launching nothing. All of us were singularly committed to exactly one thing: our mission.

Those people in that garage and that urgent need to prove our vision formed the core of the hard-driving, mission-driven culture that defined Nest.

Growing that team the right way—breaking down who we needed, how to hire them, how to build team processes and ways of thinking—was just as important as building the right product.

We borrowed some structures and norms from companies and cultures we liked, and the rest we created from scratch. We figured it out as we went, tweaked and adjusted, until we created teams and cultures that together could create something amazing.

So if you're trying to build a team, figuring out how and who to hire, then here's what I've learned about some of the key teams and competencies of most startups:

Design
Marketing
Product management
Sales
Legal

And what I learned when those teams grew. And grew. And grew.

Chapter

5.1

HIRING

A near-perfect team is made up of smart, passionate, imperfect people who complement one another. As that team grows beyond ten, twenty, fifty people, you'll need:

» Eager new grads and interns to learn from your experienced, well-seasoned crew. Every young person you spend time training is an investment in the long-term health of your company.
» A defined hiring process that ensures that candidates interview with people from across the company who they'll work with directly.
» A thoughtful approach to growth to avoid watering down your culture.
» Processes that ensure new employees are immersed in and build on your culture from day one.
» A way to keep HR and hiring top of mind for your leadership team and the management teams under them. It should be the first topic of business in every team meeting.

You'll also need to fire people. Don't be scared of it, but don't be callous, either. Give people plenty of warning and opportunity to course correct, follow the letter of the law, then bite the bullet and help them find a better opportunity.

••

One of the first people who joined Nest after me and Matt was Isabel Guenette. She was twenty-two years old, fresh out of college, brilliant and empathetic and immensely kind, ready to change the world. We hired her because we needed help researching the endless list of things we didn't know: What are the hundreds of heating systems that exist in the United States? What wires do most people have in their walls?

It didn't matter that she didn't know how to build thermostats—none of us knew how to build thermostats. That was the point. We needed to learn. So Isabel dove in.

She learned so much so fast that she became the product manager for the thermostat and successfully launched three versions over the course of five years. [See also: Chapter 5.5: The Point of PMs.]

Isabel made herself successful because she's smart and curious and capable. But some of her success also came from the fact that she was young. She may not have realized how difficult the task was before her—she just did it. And she did it with joy.

The best teams are multigenerational—Nest employed twenty-year-olds and seventy-year-olds. Experienced people have a wealth of wisdom that they can pass on to the next generation and young people can push back against long-held assumptions. They can often see the opportunity that lies in accomplishing difficult things, while experienced people see only the difficulty.

And they can grow with your company. The tried-and-true employees who joined your business in the beginning will leave eventually. Everyone leaves eventually. But before they go, you want them to mentor and train an army of young people. That's how you keep your company going. That's how you create a legacy.

You don't want to look around ten years after you launch and realize there's nobody under thirty-five working at your business.

Nest policy was always to hire a mix of new grads and to host an intern program. It was not a popular policy—not at first. The hiring managers bitched and moaned. They wanted to hire people with a ton of experience, dump a pile of work on their heads, and just let them dig themselves out.

And there's a place for that. There should always be someone on the team—or many someones—who's done it before and can do it again.

But if you look at a promising young kid or enthusiastic career switcher and see only how much time they'll take to train or the chance that they won't work out, then you're forgetting the power and drive of an ambitious talent right on the cusp of figuring out who they're going to be.

Someone took a risk on you once. Someone guided you through your mistakes, took the time to help you grow. Not only is it your duty to create that moment for the next generation, but it's also a good investment in the long-term success of your company.

Out of every ten interns we brought on each year, one to three would get offers to return again next summer or would just be hired full-time.

Even those we didn't ask back worked on real stuff, shipped real features, and made it closer to knowing what they wanted to do. Some even changed majors when they figured out what their initial career choice meant for their future. And that's what they told their friends. Suddenly over a few summers we had a pipeline of all the brilliant rising talent from the world's best universities.

That's when the hiring managers stopped bitching.

It is a battle to find amazing talent. You cannot afford to ignore any part of the population when you're trying to grow your team—there are pools of outstanding young and old and female and male and trans and nonbinary and Black and Latino and Asian and Southeast Asian and Middle Eastern and European and Indigenous people in the world who can have a profound impact on your company.

Different people think differently and every new perspective, background, and experience you bring into the business improves the business. It deepens your understanding of your customers. It illuminates part of the world that you were blind to before. It creates opportunities.

Hiring a diverse and talented team is so incredibly crucial to your success that you'll want to interview every person who joins your company yourself. But you can't. You only have twenty-four hours in a day. The seed crystals will work for only so long. [See also: Chapter 4.2: Are You Ready?: Seed crystals.] And eventually you'll have to trust the team to make its own choices.

But that doesn't mean hiring should be a free-for-all. You need a process. And the ones I've seen don't cut it.

Companies usually follow one of two methods for hiring:

1. Old school—The hiring manager finds a candidate, sets up interviews with a few people on their team, then hires the candidate. Simple. Straightforward. Stupid.
2. New school—The decision for whether to hire someone gets distributed across a ton of (typically random) employees and a fancy recruitment tool. So a candidate interviews with a bunch of people and those people enter their feedback into an evaluation form, the recruitment tool spits out a summary, then the hiring manager can bring on the candidate if they hit all the metrics. Idealistic. Novel. Stupid.

The old-school method ignores too many people at the company. The new-school method involves people who don't have enough context to make a thoughtful decision and burns people out. As you grow and can no longer rely on referrals from existing employees, you may need to bring in fifteen candidates to fill a single slot. Ask too many people to take on the burden of interviewing those candidates and they'll start to resent it, detest it, and do the bare minimum to fill out the evaluation form and get back to work.

The key is to get the candidate talking to the right people.

Nobody works in a vacuum. Everybody has internal customers—people they need to deliver to. App designers, for example, create designs for engineers to implement. In this instance, engineers are their customers. So if you're hiring an app designer, you'd better make sure they interview with an engineer.

That was the system we had at Nest. We called it the Three Crowns. Here's how it worked:

1. Crown 1 was the hiring manager. They got the role approved and found the candidates.
2. Crowns 2 and 3 were managers of the candidate's internal customers. They picked one or two people from their team to interview the candidate.
3. Feedback was collected, shared, and discussed, then the Three Crowns met to decide who to hire.
4. Matt or I would watch over it all and make the final call in the rare instance when the Crowns couldn't agree. Typically the answer if we had to get involved was no, thank you: PASS.

Even when we accepted a candidate, there was always an awareness that nobody is perfect. There were always critiques, challenges. So it was the hiring manager's job to understand potential issues from the outset, talk them through with leadership and the candidate, and commit to coaching their new team member through those challenges.

There was no mystery, no black box. Everything was on the record. Everyone knew what to expect.

Then we committed. We hired them. And despite any concerns, any potential areas for improvement, everyone started with 100 percent trust. Once you assess someone thoroughly, check references, and decide to hire them—you also have to decide to trust them. You can't start with zero trust and expect someone to prove themselves to you.

Whenever you're embarking on any journey—a new employee, a new job, a new partnership—you have to believe it'll work. Believe that people will do right by you. Of course, there will be disappointments—some people will knock your trust down to 90 percent, to 50 percent, to zero—but if you let that keep you from trusting others, you'll never know the relationships and opportunities you missed.

You can't afford that. Hiring is too important. You're going to need all the help you can get.

That's why it's also essential to have great recruiters—ones who are as excited about the company and the product as you are.

Our first recruiter at Nest was Jose Cong. We knew we had to have Jose—he gave us recruiting superpowers on the iPod and iPhone teams. How could Nest not have him, too? And the thing that sets Jose apart is twofold—he has a great eye for talent, and he is incredibly, unflappably, immensely enthusiastic. That enthusiasm is infectious and—most critically—honest. He was 100 percent sure that Nest would change the world and he'd tell the "why"—the story of the company—with the kind of zeal and joy that genuinely inspired and excited candidates. [See also: Chapter 3.2: Why Storytelling.]

Jose would bring amazing candidate after amazing candidate through our door. And then it was up to us—we had to figure out if they were right for the team. We had to interview them.

So we set up some ground rules. Everyone on the team knew what we interviewed for and what we cared about so they could calibrate on more or less the same things. We expected candidates to be mission-driven and good on their feet, the right fit for the culture, and passionate about the customer. We also had a "no assholes" policy. Pretty self-explanatory, but very helpful. If someone walked in with a ton of experience and was exactly what we were looking for on paper, but came off as unbearably arrogant or dismissive or controlling or political, then that résumé would be tossed out.

Of course, to figure out if someone you interview is an asshole, you have to know how to interview them.

This will come as a shock to absolutely no one, but I am not the world's most easygoing interviewer. I really dig in, try to understand a candidate's psyche, maybe even stress them out a little to see how they deal with stress. Everyone has different styles, but you can't be so low-key that you never dip below the surface, never push to understand who this person really is. An interview isn't about carefree small talk. You're there for a reason.

In an interview I'm always most interested in three basic things: who they are, what they've done, and why they did it. I usually start with the most important questions: "What are you curious about? What do you want to learn?"

I also ask, "Why did you leave your last job?" Not the most original question, but the answer matters. I'm looking for a crisp, clear story. If they complain about a bad manager or being the victim of politics, I ask what they did about it. Why didn't they fight harder? And did they leave a mess behind them? What did they do to make sure they left in the right way? [See also: Chapter 2.4: I Quit.]

And why do they want to join this company? That reason had better be completely different from why they left their previous job. They should have a new story, a compelling story, about what they're excited about, who they want to work with, and how they want to grow and develop.

Another good interview technique is to simulate work—instead of asking them how they work, just work with them. Pick a problem and try to solve it together. Choose a subject that both of you are familiar with but neither is an expert in—if you pick a problem in their domain they'll always sound smart; pick a problem in yours and you'll always know better. But the subject doesn't matter nearly as much as the process of watching them think. Get on the whiteboard, draw it out. What kinds of questions do they ask? What approaches do they suggest? Do they ask about the customer? Do they seem empathetic or oblivious?

You're not just interviewing to see if a person can do the job required of them today. You're trying to understand if they have the innate tools to think through the problems and jobs you don't see coming yet—the jobs they can grow into tomorrow.

Startups are always evolving, as are the people inside them. Knowing that and trusting the team and building a real hiring process allowed Nest to grow to 100, 200, 700 people.

But we were careful not to grow too fast. We wanted to hold on to our starting-team DNA—the urgency and focus of that small group in the garage, wobbling around on those terrible chairs. And the only way to do that was to integrate new people into the culture at a reasonable pace, so they could learn by doing, by watching, by working with the team and absorbing the culture organically. The best way to share and embed cultural DNA is person to person. When you're growing fast, the new people you just hired most likely have some responsibility to hire as well, so a week of orientation isn't going to cut it.

If you have fifty people who understand your culture and add a hundred who don't, you will lose that culture. It's just math.

So when bringing in new employees—especially execs—you shouldn't just throw them in the deep end, hand them a branded company notebook, and think you're done. The first month or two are crucial and should be a period of positive micromanagement. Don't worry about getting too in the weeds or not giving them enough freedom. Not at first. A brand-new person needs all the help they can get to become really well integrated. Explain how you do things in detail so they don't make mistakes and alienate the rest of the team right off the bat. Talk to them about what's working and what isn't, what you would do in their position, what's encouraged and what's verboten, who to ask for help and who to treat with kid gloves.

That's the best way to immerse someone in the culture, style, and processes of a team. Give them the push they need to start running with the pack rather than leaving them standing on the starting line, reading some docs, hoping they'll catch up.

Always remember that it's scary joining a new team. Not knowing anyone. Not knowing if you'll fit in. Not knowing if you'll succeed.

That's why I started doing brown-bag lunches with the CEO. Matt did them too. Every two to four weeks, we'd gather a crew of 15–25 new hires and existing employees and have an informal lunch. We tried to cross-pollinate different people from different groups, a good mix from around the company. No managers, no executives, no keynote presentations. Just an opportunity for them to get to know the bogeyman at the top and for me to get to know them. They asked me about our products, our policies, about me and Matt and our history at Apple. About why we didn't allow massages, about why we had so many code names. [See also: Chapter 6.4: Fuck Massages.] And I asked about what they were excited about, what they were working on, why they joined.

It was my chance to highlight why their role was important, to talk about how their team's goals powered our company goals, about our culture and our products and new projects and what was going right and what wasn't. New employees had the chance to come directly to me with their questions as well as meet existing employees who were already steeped in our culture, who could help them and lead by example.

Any employee could come to five lunches a year. And each lunch was a cultural inoculation, a vaccine against indifference and apathy, against thinking that what you do doesn't matter and that nobody at the top knows who you are.

And so we grew. Teams branched off and individuated. Individual contributors became managers. Managers became directors.

Many people rose to the challenge. Many exceeded all expectations. And some didn't. Sometimes it will turn out that the people you hired early aren't right for the team as you grow. Or sometimes you hired the wrong people right out of the gate. Or you hired mediocre people. Or you hired people who weren't a good fit for your culture, despite being spectacular otherwise.

Sometimes you hire people who just won't be successful at your company.

And then you need to fire them.

But it's important to remember that while the moment of conflict is always miserable, that moment is brief and it's your job not to fixate on it, not to dwell in it for too long. You have to quickly move from "this isn't working" to "now I'm going to do everything I can to help you find a job you like that's better for you." It's counterintuitive, but firing someone from a job they're failing at and utterly unsuited to can be a surprisingly positive experience. I've never fired someone where it didn't end up being better for both them and the company.

Sometimes life is the process of elimination. Sometimes getting fired can be a good thing. But the one thing it should never be is a surprise (unless they committed a crime; and you'd be surprised—I've seen that multiple times in my career).

Under normal circumstances nobody should ever be shocked that they're getting fired or have to ask why it's happening. They may not agree, of course. But anyone who's struggling should be having weekly or twice-monthly 1:1 meetings about that struggle. That's where issues are honestly discussed, solutions are attempted, and there's a follow-up about what worked and what didn't and what's going to happen next.

Just as people make a commitment to your company when they join it, you make a commitment to them. If you're leading a company or a large org, it is your responsibility to help people identify their challenge areas and give them space and coaching to get better or help them to find a spot at the company where they can be successful.

But even with all the goodwill and good intentions in the world, sometimes it'll become obvious to you and to the person on their way out that their issues are unsolvable, the team has lost confidence in them, and the world is full of other wonderful opportunities, with other, much less miserable jobs that you will happily help them find. And that's when they'll leave, usually of their own accord.

The process may take a month. Or two. Or three. But usually it ends amicably and everyone is better for it.

Then again, sometimes you realize you hired an asshole.

An asshole at a tiny startup can be the end of the startup. But assholes can ruin teams and products at any stage of growth, at any size company. The bigger the team, the easier it is to sneak in and start poisoning the well.

If you're managing a petty, untrustworthy tyrant, then the knee-jerk reaction will be to cut that cancer out as fast as possible. But you'll still have to take your time—tell them the situation, give them an opportunity to turn it around. The rules for firing people differ depending on where you live, so it's important to understand them and follow them exactly. Many people will happily sue you if they believe they've been fired incorrectly. Many people you thought would be great can end up dragging your whole organization down.

That's one of the most painful things about growth—in the beginning you have this incredible core of people who you know you can climb mountains with. But that phase doesn't last forever—eventually you need to add more and more and more people to the team. Sometimes you'll screw up and hire assholes or people who just can't perform or adapt to the culture. But more often the real shock of growth is that over time you'll bring on people who are just okay. Relative to the amazing people you brought in early, they'll seem unimpressive. Mostly fine, good team players, get the job done.

And that's not the end of the world. As the company expands, you need all kinds of people at all kinds of levels.

You can't wait for the perfect A+ candidate to appear for every single empty slot. You need to hire. The best of the best don't always want to join a big team, or they're tied up in another job, or you can't afford them or give them the titles or responsibilities they want.

And sometimes the people you don't expect to be amazing—the ones you thought were Bs and B+s—turn out to completely rock your world. They hold your team together by being dependable and flexible and great mentors and teammates. They're modest and kind and just quietly do good work. They're a different type of "rock star."

By far the hardest part of growth is finding the best people—in all their different incarnations—trusting your team to hire them, then making sure they're happy and thriving.

So don't flinch away from it. Make it your first order of business. Make it everyone's priority.

At many companies I've seen HR topics left to the end of team meetings, or lumped into a separate HR or recruiting meeting. But your priority is your team, its health and growth. The best way to show that is to make it the first agenda item each and every week.

Every Monday morning at Nest, that's how my management meetings started: Who are the great people we want to hire? Are we making our hiring goals or retention metrics? If not, what's the problem? What are the roadblocks? And how is the team doing? What issues do people have? How are performance reviews going? Who needs a bonus? How are we going to celebrate these accomplishments so the team feels valued? And, most importantly, are people leaving? Why? How are we going to make this job more meaningful and fulfilling and exciting than anything else out there? How are we going to help our people grow?

Only after we got through this important subject could we move on to anything else—like what the hell we were building.

The managers on the team saw it was important to me, so that's how they started structuring their weekly meetings with their teams. It became the Nest way. People first. Always.

What you're building never matters as much as who you're building it with.

Chapter 5.2

BREAKPOINTS

Growth will break your company. As more people join, your organizational design and communication style need to keep up or you risk alienating the team and cratering your culture.

Breakpoints almost always come when you need to add new layers of management, inevitably leading to communication problems, confusion, and slowdowns. In the early days of a company, when most people are self-managing, the absolute maximum number of people one human being can effectively manage directly is 8–15 full-time employees. As the company grows, that number shrinks to around 7–8. When teams approach that point, you need to preemptively create a management layer, ideally by promoting from within, and then put systems in place to ensure effective and efficient communication.

In order to keep breakpoints from actually breaking your company and causing employees to flee en masse, put management changes in place early, talk to the team about the new plan, and mentor them as they shift into different roles.

••

If you have a team of six, then six days of the year it's someone's birthday.

So you get a cake, take the afternoon to celebrate. It's nice.

When you have a team of three hundred, there's a birthday practically every day. Should we still celebrate each one? The whole team can't keep taking the afternoon off. And do you still get cake? Is cake important to your culture? You want to do everything you can for your team but there are hard realities. There are deadlines. There are budgets. And people are expensing a lot of goddamn cake.

The cake is a microcosm of the larger problems of growth, but I'm also speaking literally. Turns out people get weirdly defensive about cake. It's always a mini-crisis when you have to stop having all-company birthday parties for individual employees.

Growth can catch you off-guard like that. Because everything always falls apart just when it feels like nothing can stop you. Breakpoints typically come at moments when things are going great—business is booming or at least product development is humming along. It seems like you've finally figured it out and are on your way.

But it's kind of like having kids. Just when you think you've got a handle on things—they're eating! they're sleeping! they're walking (and getting into everything)!—your kids grow. That phase is over. Walking is old news. And everything that was working up until now utterly fails you.

It always happens. Always. And the only thing you can do is embrace it.

I've had many conversations with entrepreneurs who told me they hate it when companies grow past 120 people, so they won't let that happen in their own startup. But I've never seen it work—not for any successful business.

It's either grow or die. Stasis is stagnation. Change is the only option.

But that doesn't make it any easier.

Breakpoints happen in the transitions between team sizes. Whether we're talking about independent businesses or teams within a larger company, shifting between these size groups is always hard:

UP TO 15–16 PEOPLE

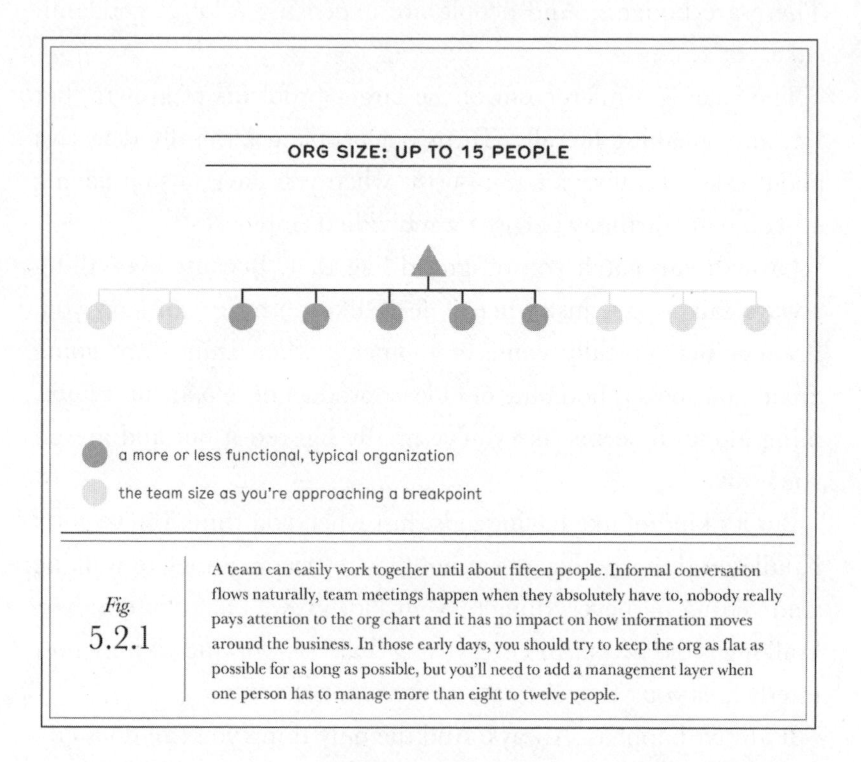

Fig. 5.2.1 A team can easily work together until about fifteen people. Informal conversation flows naturally, team meetings happen when they absolutely have to, nobody really pays attention to the org chart and it has no impact on how information moves around the business. In those early days, you should try to keep the org as flat as possible for as long as possible, but you'll need to add a management layer when one person has to manage more than eight to twelve people.

Organization: Everyone does a bit of everything and almost all decisions, major and minor, are made together. There's no need for management because the team leader helps drive the vision and decisions, but more or less acts like a peer.

Communication: Happens naturally. Everyone is in the same room (or chat room), most likely hearing all the same conversations, so there are no information bottlenecks or need for regular meetings.

UP TO 40–50 PEOPLE

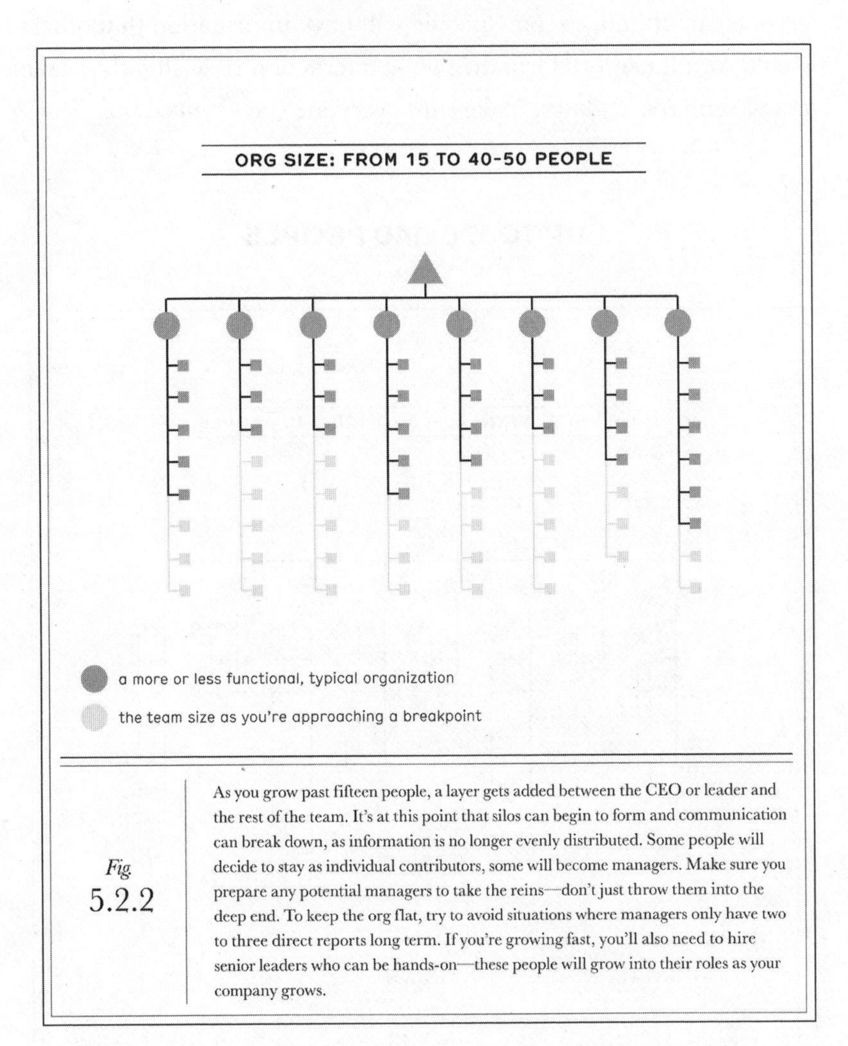

Fig. 5.2.2

As you grow past fifteen people, a layer gets added between the CEO or leader and the rest of the team. It's at this point that silos can begin to form and communication can break down, as information is no longer evenly distributed. Some people will decide to stay as individual contributors, some will become managers. Make sure you prepare any potential managers to take the reins—don't just throw them into the deep end. To keep the org flat, try to avoid situations where managers only have two to three direct reports long term. If you're growing fast, you'll also need to hire senior leaders who can be hands-on—these people will grow into their roles as your company grows.

Organization: When you go beyond 15–16 people, sub-teams of up to 7–10 people begin to form. Some people from your original core group will have to narrow their responsibilities and start managing, but the team is still so small that everything remains pretty flexible and informal.

Communication: For the first time, you'll have meetings that not everyone can attend, so some people will have information that others won't. You'll need to formalize your interaction style slightly—take notes, send out updates, make sure everyone stays synced up.

UP TO 120-140 PEOPLE

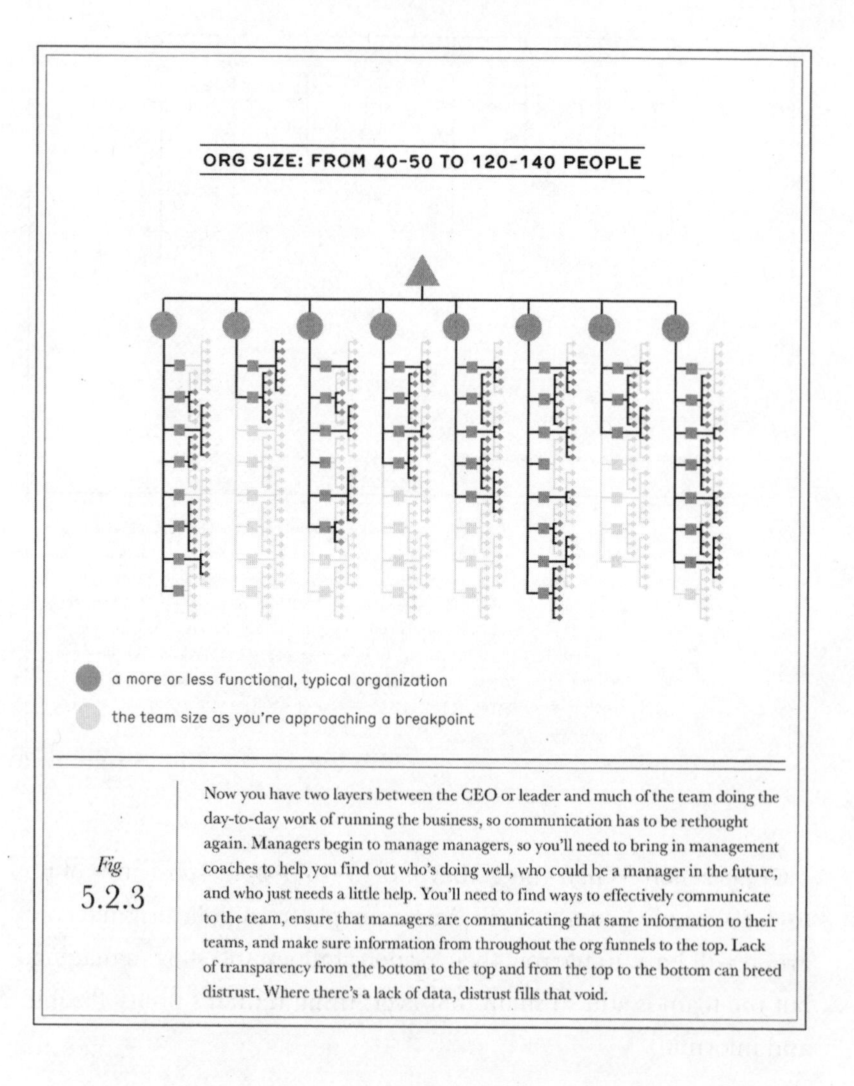

Fig. 5.2.3

Now you have two layers between the CEO or leader and much of the team doing the day-to-day work of running the business, so communication has to be rethought again. Managers begin to manage managers, so you'll need to bring in management coaches to help you find out who's doing well, who could be a manager in the future, and who just needs a little help. You'll need to find ways to effectively communicate to the team, ensure that managers are communicating that same information to their teams, and make sure information from throughout the org funnels to the top. Lack of transparency from the bottom to the top and from the top to the bottom can breed distrust. Where there's a lack of data, distrust fills that void.

Organization: As you grow past 50 people, some people become managers of managers—which is a very different beast than just managing individual contributors—and HR is really coming into play for the first time. You need proper processes to deal with promotions, defining job duties and hierarchical levels, and benefits. You'll also really need to figure out job titles.

Functional teams grow, and sub-teams within the larger teams form. Each team begins to develop its own work style around the types of work they do. Specialization is more and more necessary. Many team members begin to pick a lane and focus on a particular area, rather than have the (double-edged) luxury of being a jack-of-all-trades.

Communication: Inter-team communication has to be formalized, as do meetings with leadership. Hallway conversations won't cut it anymore. You need regular all-hands meetings where teams keep each other in the loop and execs are called upon to unite, inform, and inspire.

The execs of the company really have to nail down their communication style at this point: how do you connect with your leadership team and set priorities, how do you run meetings, how do you present to the entire company? Leadership starts to meet with HR weekly to manage the explosion of people issues.

UP TO 350–400 PEOPLE

Organization: At this point you may have multiple projects competing for the same resources. Leadership is much more isolated and distant from the actual product and spends most of its time managing org charts and conflicting priorities between teams.

Communication: Meetings are probably getting out of control and information is getting bottlenecked. You'll need to restructure your meetings and rethink your communication style. All-hands meetings will be rarer and all about reinforcing the vision for the company rather than distributing tactical information, which means there need to be other ways for people to easily access and disseminate relevant info.

In today's all-remote world, all of this is still true. And it's even more important. When the watercooler disappears and spontaneous, unstructured communication disappears along with it, you have to be even more thoughtful, disciplined, and intentional about your communication strategies. You have to give people a road map to connect with each other.

And always remember that growth is not a step function. It's not like you do just fine at 119 people but at 120 everything falls apart. You have to start forming a strategy for how you'll grow past a breakpoint long before you reach it—at a minimum two to three months before the break and then months of follow-up after. Think through your org design, your communication styles, figure out if you'll need to train individual contributors to become managers or bring in new blood, adjust your meetings, see if people scale or not. And you'll have to talk to people. A lot.

Consistency is key. If you're leading a project at a big company or running your own startup, you are going to have to coach the entire team through these transitions. The company is going through puberty, and a few awkward but critical conversations need to happen well before the first rogue armpit hair is spotted. You can even use the same language: this happens to every growing, thriving business. It's natural. Don't worry.

But you'll also need to speak openly about how it's scary for them and for you and for the company. Acknowledge that there are things that you'll lose, and that those losses will not be easy. Involve managers and individuals in the process so it doesn't just come out of left field, like something that was done to them that they have no control over. You need their help to get it right so they can define, own, and embrace the changes.

If you can see it coming, you can design your future.

But first you have to get past the fear. So here's what's going to scare people most and how to help them through it.

SPECIALIZATION

Every organism starts as a single cell. That cell divides into two, four, sixteen. In the beginning each cell is the same, but quickly they break out and individualize. This one is going to be a nerve, this one a muscle. The more the organism grows, the more differentiated each cell has to be, the more complicated the system becomes. But it also becomes more resilient, able to survive for years, decades.

The same happens in business. But people are not stem cells. Sometimes you'll work with a specialist who's thrilled by the idea of focusing on just one element of their job, but for most people narrowing their responsibility doesn't feel natural and inevitable—it freaks them out. And this process is particularly terrifying in the very beginning, after everyone gets used to doing everything, when there are virtually no management layers and you all just agree on a direction and start sprinting. But it happens later as well—even at big companies. Even at huge ones.

The fear is that everyone used to do all these cool, different things and now someone is going to come and take them away.

So focus people's attention on the opportunity: help them get curious about what their job could become, instead of being fearful of what they might lose. Do they want to be a manager? A team lead? Do they want to learn more about some other area of the business or dig deeper into something they really enjoy? What do they want to learn?

The first steps are to help them understand what they truly love about their jobs, the company, and the culture. Then they can work with their managers to retain those things and unload the stuff they don't actually like. Or they can use this moment to start something completely new.

Just keep reminding everyone that this is their opportunity to choose their path. They're in charge of their career. Tell them: project yourself into the future and figure out who you want to become and what you want to do.

ORG DESIGN

Just as people need to specialize as a company grows, so do teams. When you have only one product, you can organize by function: one hardware engineering team, a single software engineering team, etc. But as you add more product lines, that organization slows you down. It could break with just two products or it might break at five. But it will break eventually.

The problem is usually the people at the top—the team leaders can keep only so many projects in their heads. They can focus on three, four, five projects but by the time they get to six or seven their brains are fried. There just aren't enough hours in the day. So those projects get sidelined for later, and later never comes.

You'll need to break your org into product specific groups so that each product gets the attention it deserves. This team works on the thermostat, this team works on the smoke alarm. And then you might have to subdivide again. At Nest we ended up creating a team for accessories; otherwise they'd have never gotten made. The mainline team would always say they'd get to it but the accessories projects were never their first priority, so they'd inevitably prioritize other things.

Amazon, Square, Stripe, Twilio—pretty much every team with multiple product lines has had to re-org this way.

Each product family gets a dedicated engineering team, a dedicated marketing person, a dedicated designer and writer. And that turns them into little startups inside the business—smaller, faster, more autonomous. Decisions speed up and everyone shares one clear goal rather than scrambling for resources on a side project.

It just works better. But that's not to say that people will be happy about it. Groups don't enjoy being narrowed any more than individual contributors do.

But the conversation you had with individual contributors about their personal trajectory can be used again as the group trajectory narrows. Breaking things up this way flattens the org, eliminates a ton of overhead, and creates more opportunity for growth, more chances to dig in and find something you can excel at and be recognized for.

In any case, people can always switch teams. Launch another version of the old thing, then switch over to the shiny new thing. Try the thermostat, then the smoke alarm. If someone's engaged and excited, there's always wiggle room.

GOING FROM INDIVIDUAL CONTRIBUTORS TO MANAGERS

Many times a star performer will be asked to lead a new growing team. [See also: Chapter 2.1: Just Managing.] Some people will embrace the idea of management, but others will recoil in horror. It may be fear of change. It may be insecurity. Or it may be that they just really like their job and the company the way it is. In these moments, help people understand the necessity of a management layer—the team has just gotten too big, we need to specialize, we need to prepare for more growth—then give them their options:

1. **Stay an individual contributor but get layered under someone else.** This is not necessarily a bad thing; their new manager could be

a friend they've worked with for a long time at this company. Or you may be able to bring in an amazing leader from outside who they can learn from. But if they choose this option then they have to accept that they'll be managed differently and won't have as much influence over how the team evolves.

2. **Do a management trial.** Let them try out the role and see how they feel. Go on vacation and hand over the reins. Tell everyone that this person is in charge. Or start bringing them to management meetings with you—ask them to present. Ask them to lead larger and larger projects. Delegate some tasks to them, let them see what the job is really about. Have them help with HR details. Bring them into planning meetings.

    Then ask the candidate if they'd like a real trial run. Send them to manager training. If your company is too small to have proper training, then assign them an experienced manager as a coach (this should be formalized and one of the coach's Objectives and Key Results or OKRs for the quarter. It should be a key goal rather than a hand-wavy "Help this person out, would you?").

    Then go to the rest of the team in 1:1s and mention that you're thinking of promoting this person, but you want to make sure everyone is comfortable first. Say, "Let's give it a try. If you have any issues, come to me." Start getting everyone used to the idea and give the candidate time to shine.

    Then give them the option to make it real after they've gained some confidence in their abilities and the team feels comfortable with them in a new role.

Just start doing manager training early and make sure they have other experienced managers to talk to. Get them interested in the craft and science of being a good manager and explain that one of the critical parts of management is helping the team come up with creative solutions to difficult problems. You may not be doing all the work yourself, but you'll be a vital part of making it great.

I've seen a lot of people with nascent leadership talent rise to the occasion. But you should know that some won't level up. Some people will implode. Some will quit. Some will hate the job. Some will just be mediocre. In those moments, it's your responsibility to help them find other opportunities—within the company or outside it. They tried something and failed, and that means they learned. That's okay. Life is the process of elimination and now they're free to try something new.

GOING FROM A MANAGER OF ICS TO A MANAGER OF MANAGERS

At around 120 people, you need directors: managers who manage other managers. Directors need to think more like a CEO than an individual contributor.

They have to put a lot more trust in the team below them—delegate more responsibilities—while taking on the role of coach. They're close to the team but further from the product, responsible for big strategic swings but not entirely independent. And at the end of the day, they still need to deliver.

So these new directors shouldn't just be dumped into the job with no support. They should be trained and assigned coaches from the outset. Maybe that's you, maybe it's someone else, but formalize the relationship. Help the new director realize that nobody expects them to know everything right away.

MEETINGS

The first thing most people complain about when companies grow fast is the sudden crush of meetings (and emails and messages; but mostly meetings). Team meetings, management meetings, all-hands meetings, HR meetings. To a certain extent there's no avoiding them—people have to talk to each other and group chats can only get so big before they become counterproductive. You need meetings—in person or otherwise.

But you also need to occasionally stop and reevaluate your meetings and communications processes and change things up when they're no longer an effective or efficient use of time. You might turn some meetings into status update reports and reduce the number of people who attend. But then you have to be wary of too many reports—you don't want the teams spending tons of time releasing information that nobody reads. It's a constant battle. Managers should always be paying attention to how many hours teams are sitting in meetings—both intra-team and inter-team—and working to keep those numbers under control.

All-hands meetings are a great example. These are meetings that everyone in the company attends. In the beginning, when you have fewer than 40–50 people, they probably happen weekly or every other week. They start out as informal and super-tactical gatherings. For the next hour we're going to sit on the floor, share some lemon bars, discuss what everyone needs to know this week to do their jobs, address the next milestone, talk about what fun stuff we're doing, and check out the competition. Sometimes, if necessary, you deliver hard news. But usually you're looking forward—you talk about the mission and your progress toward it, then there's a bit of team building at the end.

But as more people join the team, it's impossible to make the meeting highly relevant for every person who attends and cover all the topics you want to cover. So you start to have all-hands meetings less frequently. And the content starts to change—it becomes less about what's happening right now and more about the larger vision for the company and the big swings that are being planned.

The fun, weekly, interrupt-each-other, sit on the floor covered in crumbs all-hands doesn't scale.

And if you don't recognize that, you get trapped. Like Google. Up until very recently, all 140,000 people at Google attended a 2–3 hour all-hands meeting every week—the famous (or infamous) TGIF meeting. TGIF is short for "Thank God it's Friday" but it actually happened on Thursdays since Asia needed to attend (another example of something that doesn't scale).

Aside from a bit of banter from the execs, most of TGIF was dedicated to teams from across the company presenting their work. Sometimes the content was really interesting. Many times it wasn't. But the purpose of the meeting—transmitting relevant information in an effective manner—was buried years ago. Most Googlers spent the entire three hours making memes about the meeting in an internal app called Memegen. And while that's great for culture and a nice way of uniting the team, there's no human on Earth who can claim it's efficient or will help anyone do their job better.

And it's expensive. Even if you don't factor in the cost of a large chunk of your company spending several hours a week making memes, there's so much prep work involved. Google had a dedicated team for TGIF—dozens of people spent hundreds of hours on these weekly productions.

So save the all-hands for when you really need them—make them special. Keep them regular but rare. And encourage smaller, interteam groups to get together to share relevant information. They can even sit on the floor eating lemon bars. But the goals of the meetings should be crisp and clean, and the time people spend at work should have a purpose.

HUMAN RESOURCES/PEOPLE

In the beginning, you don't really need HR. When you're five, ten, even fifty people, you can just use an external recruiter to grow the team, talk to each other when issues come up, and outsource the basics—health care, 401(k)s, etc.

But when you hit 60–80 employees, you need to bring HR in-house. Because you're not just dealing with 60–80 people. It's actually 240. Or 320. Most employees come with a family—spouses, partners, dependents. And each of those people will have some need that falls on your shoulders; they'll get sick or pregnant or need braces or want to take a leave of absence or just have questions about benefits.

It will get more and more expensive to outsource HR and will absorb way, way too much of your time.

So bring it in-house and remind employees that HR is there to protect them and the culture. To help them if they have a kid. To ensure they get paid on time. To make sure they feel safe. Adding a formal HR function doesn't take anything away—it just gives them and their families a better resource.

COACHES/MENTORS

Coaching and mentorship is critical before breakpoints. Especially at the transition to 30–40 people, when managers emerge, and around 80–120, when you promote people to director.

Just remember there's a difference between coaches and mentors:

Coaches are there to help with the business. It's all about the work: this company, this job, this moment in time.

Mentors are more personal. They don't just help with people's jobs, they help with their lives, their families.

A coach helps because they know the company; a mentor helps because they know you.

The best is a combination of the two—someone who understands both worlds—a mentor/coach who can help people see the bigger picture about what the business may need as well as what they need personally.

In the early days, if you're the leader, you're the mentor. You prepare people for the big transition and coach them through it. But as the team grows, you'll need to bring in formal mentors or coaches to help with some of the load. At 120 people, you should have executive coaches who can be there to guide your leadership team through their new responsibilities as well as communication and organizational strategies.

CULTURE

Culture is the hardest thing to pinpoint and the hardest to preserve. Even in small companies, each team typically evolves its own distinctive culture. And when a treasured part of that culture disappears, it can take a lot of your employees with it.

So to preserve what you love, have your team write down the things they value most and build a plan to continue them. And remember it's not necessarily the obvious stuff that binds people to your company—it can be small things, silly things. At Nest a few members of the team started doing barbecues in the parking lot when we were really small. They were nice—everyone would relax and talk and eat. As we grew, those barbecues could have easily petered out—steak for fifteen people is very different from steak for fifty. Or five hundred. So we invested in them as a business. They became bigger and more elaborate and more expensive, but we refused to let them die. It was vital for our culture that everyone had the opportunity to just hang out—execs and employees, design and engineers, QA and IT and support. It was just a barbecue, but it was important. And it scaled a lot better than all-hands meetings.

Culture arises organically but then needs to be codified to be maintained.

So write down your company values and post them on your physical and virtual walls. Share them with new employees. Make them part of every interview with new candidates. Everyone should know what matters at your company—what defines your culture. If you don't explicitly know your values, you can't pass them on, maintain them, evolve them, or hire for them.

And make every team write down how they do things: What's the marketing process? What's the engineering process? What are the phases for how we make a product? How do we work together? It can't just be left in people's brains. People leave. New people join. If you're growing geometrically—in all directions at once—then you need a strong, stable core at the center. Your experienced employees have to be able to walk new employees through how you do what you do, or else everyone gets lost.

I've watched these breakpoints crack hundreds of companies we've invested in and I've experienced them in my own life—when I tried to grow a new group at Philips amid a sea of nearly 300,000 employees, when Apple went from 3,000 people to 80,000. And breakpoints always seem to catch people off-guard. Nobody wants to take their eyes off their booming business and thriving vision and new products to pause and consider and restructure.

Planning for a breakpoint is a lot to deal with on top of everything else going on. And it's the worst kind of planning: messy, difficult, infinitely annoying people stuff. It's always tempting to leave it for another day.

But "if it ain't broke, don't fix it" doesn't work here. When you don't prepare for breakpoints—don't warn the team, don't thoughtfully restructure the org around roles first and individuals second, don't add new managers, don't reassess your meetings and communication tools, don't give people access to training or coaches, don't actively work to preserve your culture—then the consequences are clear:

- In their quest to keep people happy, I've seen leaders build their org around existing employees instead of first figuring out what the optimal structure should be and fitting their team into those roles.

- Then roles and responsibilities overlap, there's a ton of redundancy in the upper levels, they have to invent weird new titles for people, and nobody knows what they should be working on.
- Work slows to a crawl.
- Employees complain that the culture is dead.
- People start to quit.
- Panic sets in and it can feel like a full-blown crisis.

It usually takes six to nine months to recover. Generally companies have to trim off all the new growth beyond the breakpoint and start over again and do it right. But you have to do it right. Businesses that try to ignore breakpoints either don't survive or get stuck at their current size and stagnate.

And you should know that even if you do manage everything perfectly, you'll probably still lose some employees. Good people will quit. Some will just prefer smaller companies. Some won't like the changes, even if they respect the need for them. Some will resent getting layered, despite plenty of warning and coaching. But even if it hurts to watch trusted teammates and friends leave, the losses will be manageable. It won't be a disaster. Your culture and company will survive.

And eventually, once you've reassured your employees, trained the managers, had a million 1:1s to talk through people's anxieties, codified your values and processes, and made speeches at your regular (but not too frequent) all-hands to help build and strengthen the culture, you'll have to take a minute and think about yourself.

You're probably also scared. And you should be. If not, you aren't taking it seriously enough.

Breakpoints don't just happen to the company. They happen to you. As a CEO or founder or leader of a team in a larger company, the bigger your organization grows, the more isolated you become and the further the product retreats from you. When you first started, you helped hire everyone, knew everyone, were in many if not all the meetings, were side by side with your team, building together. As the team grew past 120–150, everything changed. You started seeing faces you didn't recognize—are these people our employees? Partners? Friends who came for lunch? You don't know all the nitty-gritty details of what's going on anymore. And you can't just walk into a meeting without freaking everyone out. Why is the CEO here? What's wrong?

So when breakpoints come for you, remember how you've been reassuring the team and take your own advice: know they're coming and get ready. Talk to your mentor. Understand what your job should look like well before every transition and plan for it. And always remember that change is growth and growth is opportunity. Your company is an organism; its cells need to divide to multiply, they need to differentiate to become something new. Don't worry about what you're going to lose—think about what you're going to become.

Chapter

5.3

DESIGN FOR EVERYONE

Everything that needs to be created needs to be designed—not just products and marketing, but processes, experiences, organizations, forms, materials. At its core, designing simply means thinking through a problem and finding an elegant solution. Anyone can do that. Everyone should.

Being a good designer is more a way of thinking than a way of drawing. It's not just about making things pretty—it's about making them work better. You may not be able to create a perfectly polished prototype without a professional designer, but you can get pretty damn far on your own if you follow two core principles:

1. **Deploy design thinking:** This is a well-known strategy originated by IDEO's David Kelley that encourages you to identify your customer and their pain points, deeply understand the problem you're trying to solve, and systematically uncover ways to solve it. [See also: Reading List: *Creative Confidence: Unleashing the Creative Potential Within Us All.*]

For example, a person complains that they have too many TV remotes. Instead of immediately jumping in and combining all the remotes into one giant, ridiculously complicated remote, you should take time to understand your customer: What do they do when they sit down on the couch? What are they watching? When are they watching? Who's watching with them? What do they use each remote for and how often? Where do they keep them? What happens when they grab the wrong one?

From there, you come to understand your customer's actual problem: they get home late and don't want to switch on a bunch of lights and wake their family, so they try to turn on the TV in the dark and can never find the right remote. Okay—we can find a solution for that.

2. **Avoid habituation:** Everyone gets used to things. Life is full of tiny and enormous inconveniences that you no longer notice because your brain has simply accepted them as unchangeable reality and filtered them out.

For example, think about the little sticker that grocery stores put on produce. Instead of just eating an apple, you now have to find the sticker, peel it off, and scrape the gluey residue off with a fingernail. The first few times you encountered the sticker, you were probably annoyed. By now you barely register it.

But when you think like a designer, you stay awake to the many things in your work and life that can be better. You find opportunities to improve experiences that people long ago assumed would just always be terrible.*

••

Vocabulary gets in the way sometimes.

Design is not just a profession.

A customer is not only a person who buys something.

A product is not just a physical object or software that you sell.

You can employ design thinking for everything you do.

---

* I did a whole TED talk about habituation, if you're interested in digging in further. You can watch it online.

Imagine you're looking through your closet, getting ready for a job interview: your customer is your interviewer, your product is yourself, and you're designing your outfit for the day. Should you wear jeans? A button-down? Is the company culture formal or informal? What do you want to project about yourself? Making that decision is a design process. Getting to the best outcome requires design thinking, even if it's unconscious.

Now you get the job. Congratulations. The jeans were a good call. But the office is ten miles away and you don't have a car. Welcome to today's design process—only now the customer is yourself.

You probably won't just run out and buy a random car—you'll think through your options. Do you actually need a car? Maybe you can take the bus or get a scooter or a bike. If you do get a car, what are all the things you'll use it for? What's your budget? Should you get a hybrid or electric car? Will you be sitting in traffic? Will you park on the street or in a garage? Will you be driving family or friends or coworkers or pets? Will you take road trips on weekends?

Design thinking forces you to really understand the problem you're trying to solve. In this case, the problem isn't "I need a car to get to work." It's actually much broader: "How do I want to get around?" The product you're designing is a mobility strategy for your life.

Literally the only way to make a really good product is to dig in, analyze your customer's needs, and explore all the possible options (including the unexpected ones: maybe I can work from home, maybe I can move closer to work). There are no perfect designs. There are always constraints. But you choose the best of all the options—aesthetically, functionally, and at the necessary price point.

That's a design process. That's how I designed the iPod. That's how I design everything.

And that's what some people think is impossible for "nondesigners." [See also: Chapter 2.3: Assholes.]

I've collaborated with many designers over the years—some of them brilliant, amazing talents—but I've also come to blows with high-minded design leaders who firmly believe that design is only for designers. They think that, when faced with a tough challenge, you always need an expert to swoop in. Someone—preferably them—with a refined aesthetic sense and an impressive degree. I've watched these designers dismiss ideas that came from engineering or manufacturing based entirely on the preconception that nondesigners can't be trusted to understand customers' needs and find thoughtful solutions. If they don't come up with it then it's not a solution at all.

It drives me completely insane.

Especially because that way of thinking is infectious. I come across tons of startups that encounter a difficult design challenge and immediately think they have to hire someone else to solve it. We don't know enough, we don't have the expertise, we need someone to do this for us.

But you shouldn't outsource a problem before you try to solve it yourself, especially if solving that problem is core to the future of your business. If it's a critical function, your team needs to build the muscle to understand the process and do it themselves.

In the early days of Nest it was clear that marketing was going to be a major differentiator for us. So when I first hired Anton Oenning to lead marketing, I asked him to get to work on the packaging. Anton is a wonderfully intuitive and empathetic marketer, master storyteller, and champion of the customer experience, but he's not a designer. Nor a copywriter. Here's how he remembers it:

"In, like, week two at Nest, Tony told me to design the packaging and write the copy. 'What? Uh. Sure. Let me call some freelance designers and copywriters I've worked with.' 'No. Has to stay in-house, under wraps.' 'Ah. Huhuh. Okay. I'll get right on that.' And it turned out to be the most liberating request of, I think, my entire career."

He learned by doing. And failing. And trying again. We rewrote the packaging more than ten times and developed a messaging process and framework at the same time. [See also: Figure 5.4.1, in Chapter 5.4.] And then, after he'd built a foundational understanding of what the packaging could be and its innate limitations, after he knew the messaging in his bones, he worked with designers and copywriters to make it perfect. But none of that could have happened if he hadn't tried it himself first. He just needed a push. Usually that's all anyone who's smart and capable really needs to shine.

You may not even need anyone to draw or make aesthetic choices. For example, take naming. It's an issue all businesses face. But rather than calling in a naming or branding agency to pick a name for you, sit down and approach the problem like a designer:

Who is your customer and where will they encounter this name?

What are you trying to get your customer to think or feel about your product?

What brand attributes or product features are most important to highlight with this name?

Is this product part of a family of products or is it stand-alone?

What will the next version be called?

Should the name be evocative of a feeling or idea or a straightforward description?

Once you come up with a list, begin to use the names in context.

How does it work in a sentence?

How do you use it in print?

How do you use it graphically?

You may not come up with a name you love, but trying it yourself will allow you to, at the very least, appreciate and understand the naming process. It will give you the tools you need to work with an agency and learn from the tricks they use to reach a final suggestion.

Sometimes you really do need to hire an expert. Sometimes a

brilliant designer can build you a ladder and help your team out of the hole they've dug for themselves. But the whole time, your team should be watching and learning and asking questions so they can make their own ladders in the future.

That's how people at all levels, on every team can begin to use design thinking in their daily work—for packaging, devices, UI, website, marketing, ordering, auditory, visual, touch, smell. They'll begin to thoughtfully design everything from the process your company uses to pay its bills to how a customer can return your product.

Across the board, your team will begin to notice the holes they're standing in, shake off their habituation to the status quo, and start making things better. Rather than looking at how other companies do things—have always done things—and copying them, your team will begin to think like their customer: "Here's how I would want to return this product." Then they'll design the process from scratch, the way it should be designed:

- Ask why at every step—why is it like this now? How can it be better?
- Think like a user who has never tried this product before; dig into their mindset, their pain and challenges, their hopes and desires.
- Break it down into steps and set all the constraints up front. [See also: Chapter 3.5: Heartbeats and Handcuffs.]
- Understand and tell the story of the product. [See also: Chapter 3.2: Why Storytelling.]
- Create prototypes all along the way. [See also: Chapter 3.1: Making the Intangible Tangible.]

Not everyone can be a great designer, but everyone can think like one. Designing isn't something in your DNA that you're simply born with—it's something you learn. You can bring in coaches and teachers, classes and books to help get everyone into the right mindset. You can do it together.

Even the greatest designers in the world can't do it alone. Most people look at Apple design and say: this is the work of Steve Jobs. This is the work of Jony Ive. But that's not remotely true. It's never just one or two people pouring out their genius into a sketchbook, then handing it to some lowly employees to execute. Thousands upon thousands of people design for Apple—and it's those teams that come together and create something truly unique and wonderful.

To be a great designer you can't lock yourself in a room—you have to connect with your team, with your customer and their environment, and other teams who may have innovative ideas to bring to the table. You have to understand your customer's needs and all the different ways you can address them. You have to look at a problem from all angles. You have to get a little creative. And you have to notice the problem in the first place.

That last point doesn't sound like a big deal. But it's huge. It's the difference between a startup employee and its founder.

Most people are so habituated to the problems in their home lives or work that they no longer realize they're problems. They simply go about their day, get into bed, close their eyes, realize they left the lights on in the kitchen, groan and grump down the stairs, without ever thinking: Why is there no light switch in my bedroom that turns off all the lights in the house?

You can't solve interesting problems if you don't notice they're there.

Here's the reason I thought the iPod could really succeed: because CDs were heavy. I love music and my CD collection in those days numbered in the hundreds. Each CD was carefully packaged in its plastic sleeve, nestled beside fifty of its closest friends in one of my carrying cases. I'd DJ on weekends—just for fun at parties—and the CDs weighed more than my speakers.

Pretty much everyone in the nineties dragged their CDs with them. Pretty much everyone had a beat-up leather case that lived in their car because it was too bulky for their bag. But pretty much nobody thought about it as a problem with a solution. Everyone just assumed it was part of life—if you wanted to listen to your music, you'd need to bring your CDs.

The people who notice the problems around them—and then dream up solutions—are mostly inventors, startup founders, and kids. Young people look at the world and question it. They're not worn down by doing the same stupid thing a thousand times—they don't assume everything has to be the way it is. They ask "why?"

Keeping your brain young is key. Seeing problems that are glossed over by others is useful. Coming up with solutions to those problems, using the vocabulary and thought process of a designer, is invaluable.

Steve Jobs called it "staying a beginner." He was constantly telling us to look at what we were making with fresh eyes. We weren't designing the iPod for ourselves—we were making it for people who had never experienced digital music. People with boom boxes and Walkmen, with beat-up leather CD cases in their cars. We were trying to introduce them to a completely different way of thinking about their music. Every tiny detail mattered for those people—they could easily get stuck, frustrated when faced with something so completely new. It needed to be smooth. It needed to feel magical.

Steve wanted people to take this beautiful little object out of the box and instantly love and understand it.

But of course that was impossible. Nothing was instant. Back then all consumer electronics with a hard drive needed to be charged before you could use them. You'd get a new gadget, take it out of the box, then you'd have to plug it in for an hour before you could turn it on. It was annoying, but it was life.

Then Steve said, "We're not going to let that happen to our product."

It was typical for electronics to run for thirty minutes in the factory to make sure they worked. We ran the iPod for more than two hours. The factory slowed down. Way down. The manufacturing team complained; costs added up. But that extra time not only let us test the iPod completely, it also gave the battery time to fully charge.

Now it's de rigueur—all electronics come with a full battery. After Steve recognized the issue, so did everyone else.

It seems like a small thing, but it was meaningful. It mattered. You opened up the box and your iPod was there to meet you, ready to change your life.

Magic.

The kind of magic anyone can do.

You just have to notice the problem. And not wait around for someone to solve it for you.

Chapter
5.4

A METHOD TO THE MARKETING

Marketing does not have to be soft and hand-wavy. While good marketing is anchored in human connection and empathy, creating and implementing your marketing programs can and should be a rigorous and analytical process.

1. **Marketing cannot just be figured out at the very end.** When building a product, product management and the marketing team should be working together from the very beginning. As you build, you should continue to use marketing to evolve the story and ensure they have a voice in what the product becomes.

2. **Use marketing to prototype your product narrative.** The creative team can help you make the product narrative tangible. This should happen in parallel with product development—one should feed the other.

3. **The product is the brand.** The actual experience a customer has with your product will do far more to cement your brand in their heads than any advertising you can show them. Marketing is part of every customer touchpoint whether you realize it or not.

4. **Nothing exists in a vacuum.** You can't just make an ad and think you're done. The ad leads to a website that sends you to a store where you purchase a box that contains a guide that helps you with installation, after which you're greeted by a welcome email.

The entire experience has to be designed together, with different touchpoints explaining different parts of your messaging to create a consistent, cohesive experience.

5. **The best marketing is just telling the truth.** The ultimate job of marketing is to find the very best way to tell the true story of your product.

••

Many people think marketing is just the bit that comes at the very end of creating something—where people who had nothing to do with product development draw up a cute little ad. Like how Coca-Cola shows you happy polar bears to convince you to drink sugar water.

These are the same people who dismiss marketing as unnecessary fluff or a necessary evil. They think it's about bullshitting people, stopping at nothing to take their money. Building the product is good and clean, but then to sell it you have to get a little dirty.

But good marketing isn't bullshitting. It's not about making something up, crafting a fiction, exaggerating your product's benefits, and burying its faults.

Steve Jobs often said, "The best marketing is just telling the truth."

If the messaging rings true, then the marketing is better. You don't have to rely on bells and whistles, stunts, and dancing polar bears—you simply explain in the best way possible what you're making and why you're making it.

And you tell a story: you connect with people's emotions so they're drawn to your narrative, but you also appeal to their rational side so they can convince themselves it's the smart move to buy what you're selling. You balance what they want to hear with what they need to know. [See also: Chapter 3.2: Why Storytelling.]

To make the story feel real, to make it tangible, you need to visualize it. You need a messaging architecture:

WHY I WANT IT	WHY I NEED IT			
	WHAT'S MY PAIN		PAIN-KILLER	
I'm stuck-in-a-rut. I crave some **INSPIRATION.**	**STASIS**	I'm still in school or in my first cubicle. Maybe I'm trying to quit my job or start my own thing. But I don't know my next move.	Build helps me find that spark again and again. Everyone has to find their own spark. Build tells me where to look for it.	**SPARK**
I don't know how to start and where I should point my compass. I want some **DIRECTION.**	**RAT-RACE**	I've always done what everyone else is doing. I'm getting too comfortable competing for increasingly scarce resources.	Build helps me build a mental framework for the future and how to chart the shortest path to it.	**LEAP-FROG**
I can't relate to founders like Zuckerberg, Musk, etc. I want realistic **ADVICE** from someone who's been in my shoes.	**INCONCEIVABLE**	I want to learn from someone I can relate to, not a Harvard or Stanford drop-out.	Tony's path to Silicon Valley is relatable. He shares painful mistakes he's made along the way, so that I can avoid them altogether.	**ACTIONABLE**
Not another self-help business book! Give me a proven **STRAIGHT-SHOOTER** who says it like it is.	**TIRED**	No ivory tower. No expectation to turn around a tanker. I need small chunks that over time have big lasting impact.	Here's a guy who's built his career from the ground up. Every step is an aggressive step forward, fueled by passion and common sense.	**FRESH**

5.4.1 This is the template we created at Nest that I've now passed along to endless startups. It's been used for everything from medical diagnostic tools to sensors for shrimp farmers. Now we're using it for this book.

First you break down the pain points that your customer is feeling or has habituated away.

Each pain is a "why"—it gives your product a reason to exist.

The painkiller is the "how"—these are the features that will solve the customer's problem.

The "I want it" column explains the emotions that your customers are feeling.

The "I need it" column covers the rational reasons to buy this product.

The whole product narrative should be in there—every pain, every painkiller, every rational and emotional impulse, every insight about your customer. It needs to encompass everything because:

1. **It's essential for product development:** Product management and marketing work on the messaging architecture from day one. In order to build a great product, each pain has to be extremely well understood and answered with a painkiller in the form of a product feature. The messaging architecture is a sister text to the plain list of features and their functionality that makes up your basic product messaging. Both need to exist side by side: the what and the why.
2. **It's a living document:** As the product and your team's understanding of the customer evolve, so does the messaging architecture.
3. **It's a shared resource:** Everyone who is responsible for any customer touchpoints should be looking at this document, not just marketing. It should steer engineering, sales, and support as well. Every single team should be thinking about the what, the why, and the story you're telling.

But the messaging architecture is only the first step.

For every version of the story, we wrote down the most common objections and how we'd overcome them—what stats to use, what pages of the website to send people to, what partnerships to mention or testimonials to point to. We figured out which story we could put on a billboard all the way down to the story we'd tell a longtime customer.

The process of convincing someone to buy and use your product needs to respect the customer, needs to understand their needs at different points of the user experience. You can't just shout your top ten features at people in a billboard and a website and packaging just like you can't simply hand someone your résumé at an interview, then lunch, then on a date. Sure, you're giving them important information, but different moments in the journey require different approaches.

Your message needs to fit the customer's context. You can't say everything everywhere.

So when we were thinking about how the thermostat would reach customers, we laid out all the different ways people could discover our brand: advertising, word of mouth, social media, reviews, interviews, in-store displays, launch events.

Then we laid out the next step in the process—how they'd learn about our product. Brochures, our website, packaging, etc.

And then we created a messaging activation matrix.

When we were deciding what went where, it was crucial to know which parts of the story customers would be exposed to at various points of the journey.

- Top-line billboards would just introduce the idea of a new kind of thermostat.
- The packaging would highlight the top six features and how the product connects to your phone.
- The website would emphasize energy savings and showcase how Nest fits into your daily life.
- The usage guide inside the packaging would provide more detail about how to train the learning algorithm and tips for saving energy.

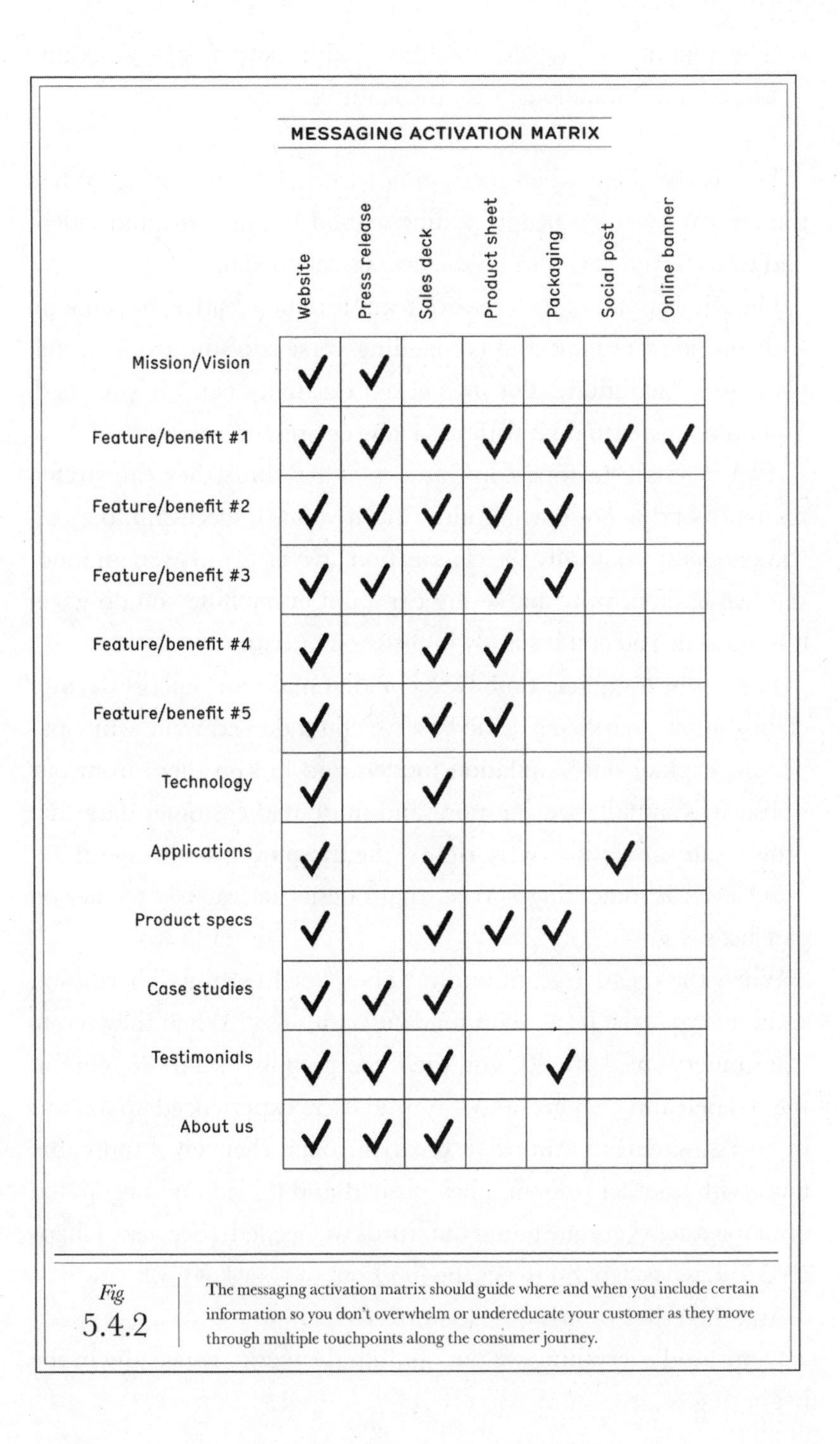

MESSAGING ACTIVATION MATRIX

	Website	Press release	Sales deck	Product sheet	Packaging	Social post	Online banner
Mission/Vision	✓	✓					
Feature/benefit #1	✓	✓	✓	✓	✓	✓	✓
Feature/benefit #2	✓	✓	✓	✓	✓		
Feature/benefit #3	✓	✓	✓	✓	✓		
Feature/benefit #4	✓		✓	✓			
Feature/benefit #5	✓		✓	✓			
Technology	✓		✓				
Applications	✓		✓			✓	
Product specs	✓		✓	✓	✓		
Case studies	✓	✓	✓				
Testimonials	✓	✓	✓		✓		
About us	✓	✓	✓				

Fig. 5.4.2 The messaging activation matrix should guide where and when you include certain information so you don't overwhelm or undereducate your customer as they move through multiple touchpoints along the consumer journey.

- The support site would go deeper with exact instructions and thorough explanations of all the features.

This is the point when messaging turned into marketing. When the facts we wanted people to understand became ads and videos and tweets. And this is when the lawyers stepped in.

The whole point of the creative team is to be creative, to come up with the most elegant and compelling version of the truth, to tell your story beautifully. But unchecked creativity can get you sued. You do not want to do it without a lawyer present.

A lot of small startups skip this step. They think they can stretch the truth and nobody will notice. But if you are successful, they always notice—especially the class-action lawyers. And even an innocent white lie in your marketing can taint everything you do when it's exposed. You can instantly lose customer trust.

That's why for a long time Nest couldn't make any energy-savings claims in our marketing—the best we could do was write white papers to explain our simulation models and link to them from our website. Eventually we got more and more real customer data that proved our simulations were right—the thermostat saved energy.

But even if something is true, that doesn't necessarily mean you can just say it.

When the creative team wrote "The Nest Learning Thermostat saves energy," the legal team made it "can save." When they wrote "Customers saved 20–50% on their energy bills," legal brought out the red pen and changed it to "Typical users experienced up to 20% in energy savings." And then creative rolled their eyes and came back with another option. They pushed and pulled and negotiated until together everyone found the words we needed. [See also: Chapter 5.7: Lawyer Up: So to get the most out of your lawyer.]

And then they brought those words to me.

I approved everything we put out into the world. Especially in the beginning.

This wasn't my area of expertise—I had watched Steve Jobs sell iPods and iPhones and work closely with marketing, but had never done it myself. So the only way I could master marketing was to bury myself in it—to take the customer journey myself, to touch every touchpoint. So nothing was ever presented to me without context—I would always expect to see what came before, what came after. I'd need to know the story we were telling and to whom we were telling it and at what point of the journey they were. An ad can't be understood without knowing where it will appear and where it will lead to. A webpage can't be approved until you know who will be routed to that page and what they'll need to know and where the call to action will take them. Everything is connected to everything, so everything must be understood together.

It wasn't micromanagement—it was care. I was putting the same amount of energy and time into the beginning of the customer journey as I put into the end. To those who aren't used to it, that feels intense and unwarranted, but that was my job. [See also: Chapter 6.1: Becoming CEO: So your job is to care.] I wanted the words and images that we used to describe our products to be as great as the products themselves. I wanted the entire experience to shine. I wanted the marketing team to be as exacting as the engineering and manufacturing teams—to learn from this rigor so they would begin to push themselves just as hard, or harder, than I pushed them.

I knew marketing was going to have to be one of our differentiators—something that lifted us beyond anything any other thermostat maker could dream of—so it was important to give it time and attention. And money.

The money is important. We were a small company with limited resources but we invested in marketing. We invested in creating beautiful things because we knew we'd amortize the hell out of them. We used every expensive, gorgeous photo in a thousand different places; we played every high-quality video everywhere we could. The team picked the elements that would have the most impact—

that we could use and reuse for years—then spent the money to do them well.

Today, ten years later, Google Nest is still using some of the photos and assets we created before we even launched our company.

The reason is that marketing was part of the process from day one. Nobody was ignoring it, nobody was forgetting about it. We knew it was useful, so we used it.

This perspective and focus allowed us to do something somewhat unique to Nest: marketing prototyped the product narrative in parallel with product development.

The clearest expression of that was the Why We Made It page on nest.com.

Fig. 5.4.3

We literally took the "why" at the heart of thermostat product development and slapped it on the front page of our website. One of the first tabs on nest.com was called "Why we made it"—that's where we connected directly with a skeptical audience, where we injected the virus of doubt. [See also: Chapter 3.2: Why Storytelling.] We explained why people thought thermostats didn't matter—why they were neglected and ignored—and then we told customers about the incredible impact they had on people's homes, their bills, and the environment.

The question of why we made it connected explicitly to why anyone should buy it. We had to get it right—for our customers and for ourselves.

It took weeks to get this page written. And as the product evolved, the page evolved alongside it. Marketing was always there, ensuring that we still had a strong answer to "Why We Made It" even as engineering and product management tweaked what we were making.

That allowed marketing to have a valuable voice in product development. Because any big change to the product would force a change to the story. It was marketing's job to figure out if that change would break the packaging, the website—all our prototypes of the product narrative. If something broke, then it was marketing's job to speak up. To talk to product management and engineering. To figure out if there was a way to make it work or if the change was untenable.

And the "Why We Made It" page was only one part of our prototype. It supplied the rational argument for why anyone should buy a Nest Learning Thermostat—because regular thermostats waste energy and wasting energy is bad for you and bad for the planet. But we needed to prototype the emotional argument as well—and for that the creative team made a video and a Living with Nest webpage that gloried in the beauty and simplicity of the product, that made it a covetable object, a piece of art on your wall to make your home better, more comfortable.

Each part of the website highlighted a different part of the product story. That forced us to know the story in our bones, to live and breathe it so we could pass it to others in the clearest, most honest way possible.

Finding the best, most honest expression of a product or feature is not easy. That's why there's an entire team devoted to it. Product management can create the messaging—the top features, the problem statement—but finding the best way to tell that story to customers is an art. It's a science. It's marketing.

Of course, that's not to say we always got it right.

There was no successful model to copy for how to sell thermostats to people who had used them a million times but had never given them a second thought. We didn't know what would connect and what wouldn't, if people would laugh at a $250 thermostat that you had to install yourself or fall in love.

So we tried things. And we screwed up plenty. I screwed up a ton.

We launched expensive brand-only ad campaigns before anyone knew our products. We made webpages so information-dense that almost nobody read them. And the customers in our imaginations weren't the customers who showed up to buy our products. They were all different, needed different things, expected different things, skimmed our laboriously written treatises in half a second and dug deep on things that never occurred to us.

But screwing it up is how we got better. It's how we learned. Brand ads made our egos feel better but they didn't drive sales (you have to make great products for years before customers will buy a product simply because of your brand). The website needed to be short and sweet and fit product info into the context of people's daily lives. Our support site needed to be more searchable because people didn't follow the paths we laid out for them.

With every piece of marketing we made, we got better at marketing. I got better at marketing. The entire company got better at marketing. The messaging architecture and activation matrix turned a soft art into a hard science that everyone could understand. And when everyone can understand it, they can understand how important it is.

Chapter

5.5

THE POINT OF PMs

The majority of companies I work with misunderstand the role of the product manager—if they even know it exists. They think it's marketing (nope), it's project management (nope), it's press relations/communications (nope), it's design (nope), product finance (nope), it's the founder or CEO's job (not really). The confusion mostly stems from the fact that product management lives at the intersection of many specialties and can look very different at different companies. But it's also because of the stupid abbreviation. A PM can refer to:

Product manager or product marketing manager—Product marketing and product management are essentially the same thing—or at least they should be. A product manager's responsibility is to figure out what the product should do and then create the spec (the description of how it will work) as well as the messaging (the facts you want customers to understand). Then they work with almost every part of the business (engineering, design, customer support, finance, sales, marketing, etc.) to get the product spec'd, built, and brought to market. They ensure that it stays true to its original intent and doesn't get watered down along the way. But, most importantly, product managers are the voice of the customer. They keep every team in check to make sure they don't lose sight of the ultimate goal—happy, satisfied customers.

Project manager—Coordinates tasks, meetings, calendars, and assets to enable individual projects to get done on time. It's important to note that project managers are more than just glorified note takers. If the product manager is the voice of the product, the project manager is the voice of the project—their job is to alert the team to potential problems that could stall or derail the project and to help find solutions.

Program manager—Supervises groups of projects and project managers, focusing on both long-term business objectives and short-term deliverables.

To complicate matters further, some companies use different titles for product managers. Microsoft, for example, calls them program managers. There are also jobs that are adjacent to product management but not quite the same, especially in the world outside technology. CPGs (consumer product groups) like Colgate-Palmolive employ brand managers who don't write the spec for the product, but are still the voice of the customer and responsible for shaping what the product will become.

In the interest of eliminating the confusion around what *PM* stands for, let's use the following abbreviations:

PdM = Product manager

PjM = Project manager

PgM = Program manager

••

When yet another CEO tells me they have no idea what a product manager does, it always makes me think of design in the eighties.

Because most tech companies in the eighties didn't have designers.

Things were obviously designed and those designs were just as critical as they are today, but nobody employed designers to architect the user experience. Designing meant making something look nice, and that just happened as you developed the product—a mechanical engineer would draw something up, or if you wanted to get fancy, you'd outsource those drawings to an agency.

There was no school for design. No formal training. And any designers who managed to get hired were second-class citizens. They didn't have the authority to push back on engineers who'd cut corners with a shrug and say, "Well, we got most of what the designer asked for. But we can't do it all—too much time, too expensive. Ship it as is!"

And then Apple, Frog, David Kelley, IDEO, and design-led thinking came along in the nineties and elevated design. Designers stopped reporting to engineering. Design schools were founded. The profession came into its own as a formal discipline—understood, respected.

Product management is on that path now. But unfortunately we're not there yet.

It's only in the last five to ten years, since the advent of the iPhone and the app economy, that certain companies have begun to really understand product management and appreciate its value. Many still don't.

It's an issue I see at a lot of startups and project teams at larger companies—the founder or team lead often plays the role of the product manager in the beginning. They define the vision and work with all parts of the business to make it a reality. The trouble comes when the team grows—to 40, 50, 100 people. [See also: Chapter 5.2: Breakpoints.] That's when the leader has to step away from the day-to-day business of building the product and hand over the reins to someone else.

But they can't imagine handing over their baby. How could anyone understand it or love it or help it grow as well as they could? And how would that function even work? Where would it live? How could the founder retain influence over the product if they're no longer the manager of that product? And then what would the founder's job even be? [See also: Chapter 6.1: Becoming CEO.]

The same thing happens at big companies. They're also flummoxed. The engineers figure out what to build or the sales team tells them what customers need. Where does product management fit in?

As we write this in 2021, Google has moved to give product managers more power for the first time. Google has always been technology and engineering-led, but now Search is being rearranged to favor product managers over engineers. It's a huge move and a dramatic cultural shift.

And the reason for it is simple: the customer needs a voice on the team. Engineers like to build products using the coolest new technology. Sales wants to build products that will make them a lot of money. But the product manager's sole focus and responsibility is to build the right products for their customers.

That's the job.

The tricky thing is that the responsibilities of a product manager are completely different at different companies. Product management is less a well-defined role and more a set of skills. It lives between everything, a white space that morphs based on the customer, the needs of the business, and the abilities of the humans involved.

A good product manager will do a little of everything and a great deal of all this:

- Spec out what the product should do and the road map for where it will go over time.
- Determine and maintain the messaging matrix.
- Work with engineering to get the product built according to spec.
- Work with design to make it intuitive and attractive to the target customer.

- Work with marketing to help them understand the technical nuances in order to develop effective creative to communicate the messaging.
- Present the product to management and get feedback from the execs.
- Work with sales and finance to make sure this product has a market and can eventually make money.
- Work with customer support to write necessary instructions, help manage problems, and take in customer requests and complaints.
- Work with PR to address public perceptions, write the mock press release, and often act as a spokesperson.

And then there's the even less well-defined stuff. Product managers look for places where the customer is unhappy. They unravel issues as they go, discovering the root of the problem and working with the team to solve it. They do whatever is necessary to move projects forward—that could be taking notes in meetings or triaging bugs or summarizing customer feedback or organizing team docs or sitting down with designers and sketching something out or meeting with engineering and digging into the code. It's different for every product.

Sometimes product managers need to be extremely technical—usually in B2B settings where the user of the product is extremely technical, too. If you're selling brake systems to a car company, you'd better really understand brakes. Having a deep well of knowledge about brakes is the only way you'll connect with your customer and understand what they care about.

But if you're building a car for a regular person, you don't need to know every detail about how the brakes work. You just need to know enough to communicate with the engineers who build them. And then you need to decide if those brakes are an important part of the marketing story you tell customers.

Most tech companies break out product management and product marketing into two separate roles: Product management defines the product and gets it built. Product marketing writes the messaging—the facts you want to communicate to customers—and gets the product sold.

But from my experience that's a grievous mistake. Those are, and should always be, one job. There should be no separation between what the product will be and how it will be explained—the story has to be utterly cohesive from the beginning.

Your messaging is your product. The story you're telling shapes the thing you're making. [See also: Chapter 3.2: Why Storytelling.]

I learned storytelling from Steve Jobs.

I learned product management from Greg Joswiak.

Joz, a fellow Wolverine, Michigander, and overall great person, has been at Apple since he left Ann Arbor in 1986 and has run product marketing for decades. And his superpower—the superpower of every truly great product manager—is empathy.

He doesn't just understand the customer. He becomes the customer. He can shake off his deep, geeky knowledge of the product and use it like a beginner, like a regular person. You'd be surprised how many product managers skip that hugely necessary step—listening to their customers, gaining insights, empathizing with their needs, then actually using the product in the real world. But for Joz, it's the only way.

So when Joz stepped into the world with his next-gen iPod to test it out, he fiddled with it like a beginner. He set aside all the tech specs—except one: battery life.

Nobody wanted their iPod to die in the middle of a flight or as they were DJing a party or going for a run. But as the product evolved from the classic iPod to the iPod Nano, we were in a constant tug-of-war—the smaller and more elegant it became, the less room there was for a battery. But what's the point of a thousand songs in your pocket if you have to keep taking them out of your pocket to re-charge?

One charge had to last days, not hours.

Battery life mattered to customers. And it mattered to Steve Jobs. You couldn't just come to Steve and say, "The next version of the iPod is going to have a twelve-hour battery instead of fifteen like the last version." You'd get thrown out of the meeting.

So Joz and I didn't bring Steve numbers—we brought him customers. Commuters like Sarah only use the iPod going to and from work, students like Tom use it throughout the day, but in short bursts between classes or basketball games.

We created typical customer personas, then walked through the moments in their life when they used their iPods—while jogging, at parties, in the car. And we showed Steve that even if the number engineering gave us was twelve hours, those twelve hours actually lasted most people all week long.

The numbers were empty without customers, the facts meaningless without context.

Joz always, always understood the context and was able to turn it into an effective narrative. It's how we were able to convince Steve. And reporters. And customers. It's how we could sell iPods.

And that's why product management has to own the messaging. The spec shows the features, the details of how a product will work, but the messaging predicts people's concerns and finds ways to mitigate them. It answers the question, "Why will customers care?" And that question has to be answered long before anyone gets to work.

Figuring out what should be built and why is the hardest part of building. And it's impossible to do it alone. Product management can't just throw a spec over the fence to the rest of the team—every part of the business should be involved. That doesn't mean the product manager should build by consensus, but engineering, marketing, finance, sales, customer support, and legal will all have ideas and useful insights that will help shape the narrative before the product is built. And they'll continue to improve that narrative as the product evolves.

A spec and messaging aren't instructions that are set in stone. They flex and change, shifting as new ideas are introduced, or new realities slap you in the face. Building a product isn't like assembling an IKEA chair. You can't just hand people instructions and walk away.

Building a product is like making a song.

The band is composed of marketing, sales, engineering, support, manufacturing, PR, legal. And the product manager is the producer—making sure everyone knows the melody, that nobody is out of tune and everyone is doing their part. They're the only person who can see and hear how all the pieces are coming together, so they can tell when there's too much bassoon or when a drum solo's going on too long, when features get out of whack or people get so caught up in their own project that they forget the big picture.

But they're also not directing everything. Their job isn't to be CEO of the product—or, God forbid, what some companies call the "product owner." They can't single-handedly dictate what will and will not make it in. Sometimes they'll have the final opinion, sometimes they'll have to say "no," sometimes they'll have to direct from the front. But that should be rare. Mostly they empower the team. They help everyone understand the context of what the customer needs, then work together to make the right choices. If a product manager is making all the decisions, then they are not a good product manager.

It's the contributions of everyone on the team that ultimately define the melody, that turn noise into a song.

But of course it doesn't always flow beautifully.

The engineers may want more say in what they're building—they may claim the product manager isn't technical enough or simply that engineering knows best. The marketers rarely want to stick to the playbook—they want to stretch and be creative, using words or images that may unintentionally misrepresent the product. People won't always get along. Opinion-driven decisions will be debated ad nauseam. Teams will get out of step, individuals will get angry, the product will get pulled in opposite directions.

So the product manager has to be a master negotiator and communicator. They have to influence people without managing them. They have to ask questions and listen and use their superpower—empathy for the customer, empathy for the team—to build bridges and mend road maps. They have to escalate if someone needs to play bad cop, but know they can't play that card too often. They have to know what to fight for and which battles should be saved for another day. They have to pop up in meetings all over the company where teams are representing their own interests—their schedules, their needs, their issues—and stand alone, advocating for the customer.

They have to tell the story of the customer, make sure everyone feels it. And that's how they move the needle.

The other day I was talking to Sophie Le Guen, an incredibly sharp and empathetic product manager at Nest.

She told me about a very early meeting she had with the engineering team to discuss the "why" for the new Nest Secure security system. For the mostly male engineering team, the "why" was simple: I want a security system to protect my home when I'm away.

But Sophie had been interviewing people and noticed that while men usually focused on empty homes, women focused on full ones. When they were alone or alone with the kids, women wanted extra protection. Especially at night.

Sophie's job was to tell their story—to help a single engineer who lived alone understand a parent's perspective. And then her job was to turn that perspective into features that worked for the entire family—a family that wanted to be safe, to turn on the security system when they walked in the door—but who didn't want to feel like prisoners in their own home. So when Nest Secure launched, the motion sensors had a single button. A homeowner (or their kid) could push the button and open a door or window from the inside without having to deactivate the whole security system or cause a blaring false alarm.

The customer story helped engineering understand the pain point. They built a product to address that pain. Then marketing crafted a narrative that gave every person who had experienced the pain a reason to buy the product.

The thread that tied all these people and teams and pains and desires together was product management. For every successful product and company, all parts of your business end up leading back to them—it's all hinged together in one central point.

This is why product managers are the hardest people to hire and train. It's why the great ones are so valuable and so beloved. Because they have to understand it all, make sense of it. And they do it alone. They're one of the most important teams at a company and one of the smallest.

Because the needs of each product and company are so different, these are incredibly difficult jobs to describe (See also: the previous three thousand words), never mind actually hire for. There's no set job description or even a proper set of requirements. Many people assume product managers have to be technical, but that's absolutely not true. Especially in B2C companies. I've met many great product managers who are able to build trust and a rapport with engineering without any kind of technical background. As long as they have a solid basic understanding of the technology and the curiosity to learn more, they can figure out how to work with engineering to get it built.

There's no four-year college degree for product management, no obvious source you can hire from. Amazing product managers usually emerge from other roles. They start in marketing or engineering or support, but because they care so deeply about the customer, they start fixing the product and working to redefine it, rather than just executing someone else's spec or messaging. And their focus on the customer doesn't cloud their understanding that ultimately this is a business—so they also dive into sales and ops, try to understand unit economics and pricing.

They create the experience they need to become great product managers.

This person is a needle in a haystack. An almost impossible combination of structured thinker and visionary leader, with incredible passion but also firm follow-through, who's a vibrant people person but fascinated by technology, an incredible communicator who can work with engineering and think through marketing and not forget the business model, the economics, profitability, PR. They have to be pushy but with a smile, to know when to hold fast and when to let one slide.

They're incredibly rare. Incredibly precious. And they can and will help your business go exactly where it needs to go.

Chapter

5.6

DEATH OF A SALES CULTURE

Salespeople are traditionally paid on commission. That means that after a customer completes a transaction, the salesperson gets paid some percentage of the sale price or receives a bounty—a bonus—for each sale they complete. The bigger the deal, the more deals they close, the bigger the paycheck. Typical commissions are paid out in full at the end of the month or quarter.

This is commonly believed to be the best way to align business goals with sales team goals and hit revenue targets that will show investors real progress. People—especially salespeople—will tell you that this is the way it's always been done, that it's the only way to do it and it's the only path to hiring a decent sales team. Those people are wrong.

Even if on the surface everything seems to be working, there are a lot of downsides when the traditional commission model is fully played out. Most notably, it can breed hypercompetition and egoism and incentivize making a quick buck rather than ensuring that customers and the business are successful in the long term.

There is a different model that aligns short-term business goals without neglecting long-term customer relationships. It's based on vested commissions.

Rather than focusing on rewarding salespeople immediately after a transaction, vest the commission over time so your sales team is incentivized to not only bring in new customers, but also work with existing customers to ensure they're happy and stay happy. Build a culture based on relationships rather than transactions.

Here's how to set it up at your company:

1. If you're starting a new sales organization, do not offer traditional monthly cash commissions. It's best to have everyone in your company compensated in the same way—so offer salespeople a competitive salary and sales performance bonuses of additional stock options that vest over time. Stock provides a built-in incentive to stick around and invest in long-term customers who are good for the business.

2. If you're trying to transition to a relationship-driven culture, you may not be able to kill traditional commissions right away. In that case, any stock or cash (stock is still preferable) that you give as a commission should vest over time. Pay 10–15 percent of the commission at first, then another tranche in a few months, then another a few months after that, etc. If the customer leaves, the salesperson loses the remainder of their commission.

3. Every sale should be a team sale. So if you have a customer success team (the team that actually delivers, sets up, and maintains whatever is sold to the customer), then it should sign off on every deal. Sales and customer success should be under one leader, in the same silo, being compensated in the same way. In this setup, sales can't just throw a customer over the fence and never think about them again. If there's no customer success team, then sales should work very closely with customer support, operations, or manufacturing—create a board of people to approve each commitment.

••

I didn't learn all this at General Magic. Or Philips. Or Apple. Or Nest.

I learned it first from my dad.

He was a salesman for Levi's in the seventies, when Levi's jeans were an international obsession. He could have made tons of money unloading Levi's crappier designs on retail stores and swiftly moving on. But he was a great salesman. Year after year he won all the sales awards—I'd see the trophies and plaques he'd bring home. And his goal was never short-term gain. It was trust.

So he showed his customers the entire lineup of products, told them which were selling well, which weren't. He steered them toward the cool styles, away from the stuff no one was buying. And if the customer wanted something he didn't offer, he pointed them to a competitor who did.

Those customers remembered him. And the next season or the next year or ten years down the road, they would call him up again. They'd make an order. And again the next season. And the season after that.

My dad was on commission but he would often sacrifice a sale in order to build a personal connection. The best salespeople are the ones who maintain relationships even if it means not making money that day.

Those are the salespeople you want on your team. Because if you do it right, they truly will become part of the team, rather than mercenaries who swoop in, make their money, then jump ship to the next hot company, leaving a trail of problems behind them.

The danger with traditional commission-based sales models is that they create two different cultures: a company culture and a sales culture. The employees in these two cultures are compensated differently, think differently, care about different things. Hopefully most people in your company will be focused on the mission—on achieving something great together, grinding away at a big, shared goal. Many salespeople won't give two shits about your mission. They'll be focused on how much they're making month to month. They'll want to close deals and get paid. It won't matter what they're selling as long as it sells.

The bigger your company, the further these two cultures will drift apart. Huge commissions, sales awards, and sales conferences where everyone high-tails it to an island, ready for a weekend of drinking, may feel great for your sales team in the moment. But they can drag morale down for the rest of the company. Why are we here working, building this thing, while they're getting wasted in Hawaii, doing shots out of their Best Salesperson of the Year trophy?

And this isn't to say sales is not important. It's vital. It brings in customers and cash that are absolutely necessary to keep a company alive. But it's not more important than engineering or marketing or ops or legal or any other part of your business. It's just one of many critical teams, all working together to make something great.

But if sales is off to the side, doing their own thing, barely part of the company but steadily meeting their monthly goals, that can breed an insulated, transactional culture. And the way customers are treated in that kind of culture can be brutal—even in places where you'd assume customers must be treated well in order for salespeople to make any money.

I worked for a commission exactly once—when I was sixteen years old, selling crystal and china at a department store called Marshall Field's. And I was great at it—the old ladies loved me. They'd pinch my fat cheeks, ask about my mother, get my address to send me Christmas cards—and leave with armloads of crystal glasses, dishware, and odd-looking fine china sculptures. It completely infuriated all the other salespeople. We were paid almost entirely in commission every two weeks and this no-nothing kid was taking food out of their mouths. So every time a nice old lady beelined toward me, they'd try to steal her and my sale. They'd literally fight over who was going to make the sale as the customer was standing right in front of us. They didn't care who this person was or what she wanted—they were only after the five- or ten-dollar commission.

And this was at Marshall friggin' Field's. That feeling gets multiplied exponentially as the dollar amounts and pressure grow. Things get much more cutthroat. And they get gross.

There are a lot of movies about terrible sales cultures—*Boiler Room*, *The Wolf of Wall Street*, *Glengarry Glen Ross*. They're sensationalized, but not by much. The hypercompetition often breeds the kind of ego-driven, boozy, locker-room backslapping where everyone ends up at a strip joint, trying to drink each other under the table. Reasonable people are trapped and feel like they have to keep up appearances, while unreasonable people spin wildly out of control—puke in hotel lobbies, get dragged out of the company holiday party by cops.

It happens all over the place—from Silicon Valley to New York to Jakarta, in tiny businesses and vast corporations. Companies think they can control the worst of it—that a little bad behavior is just the cost of a high-powered sales team. What's the problem if everyone's hitting their sales targets?

The problem is that one day something will go wrong. Maybe it'll be with the product—you'll have an issue and business will slow down. In that moment, the time when you need them most, your sales team will abandon you. They'll go wherever the sales are hot. Why should they stick with you if they can't make money right now?

Or you might find out that those great numbers they've been putting up aren't so great after all. Maybe they've been telling little white lies about your team's capacity or your product's ability to meet customers' needs. Maybe all those customers flowing into your business have been sold something you can't actually give them. And now they're pissed.

When you start out, your first customers are incredibly precious. They're the ones who love you best, who take a risk on you. And they can make or break your company—they're the source of all your initial word of mouth. In the beginning it feels like you know practically each customer by name and face and Twitter handle. But as your business grows and a traditional sales culture takes hold, those customers stop being seen as individuals. They turn into numbers. Into dollar signs.

But they're still people, even when you're in hypergrowth mode. The relationships you form with them are still meaningful and necessary. Really great, enlightened salespeople hold on to those relationships. But many salespeople won't.

If your sales culture is driven by transactions, then any relationship the salesperson cultivates will evaporate immediately after the customer signs on the dotted line. You don't have a relationship with an ATM machine—you just walk up to it and take your money. And once a customer feels like an ATM, clawing them back is almost impossible. You have to bend over backward, twisting yourself into knots trying desperately to convince them to trust you again. Customer success or support will apologize and apologize and backpedal, cursing the sales team under their breath the whole time.

You'll probably still lose that customer.

That's why relationship sales cultures aren't naive or simplistic. They're necessary. And they're proven. It's the sales culture we set at Nest. It's the culture I've pushed dozens of startups to adopt. It works better. Every single time. You get happier customers. You get a happier culture. You get teamwork and focus and progress toward your goal.

Ideally you set up your business this way from the start. Everyone gets paid in salary, stock, and performance bonuses—sales, customer success, support, marketing, engineering, everyone. That's not to say they get paid the same amount, but the compensation model is the same—everyone is aligned.

And nobody ever works a sale alone. During the sales process, the salesperson has backup from customer success or support or whoever will be working closely with the customer post-sale. And then those teams sign off on the deal. There are never any surprises—everyone knows exactly what's expected of them to make this new customer successful. And once the deal is closed, the salesperson doesn't disappear. They stay on as a point of contact for that customer, and if there's any kind of issue, they step in to resell them.

If you already have a transaction-oriented sales org and want to make the shift to a relationship-based one, it'll be trickier. People will probably leave. A lot of them will tell you you're crazy. But it can be done.

First set up a mini–internal board populated by those other teams—customer support, customer success, operations—to approve each sales deal. That will start shifting the mindset from lone-wolf salesperson to being part of a team. Then start talking about the change to commissions. Don't say you're getting rid of them—that messes with people's heads—just say that you're doing them differently. Boost the size of the commission but start vesting it over time. And tell the sales team they'll lose the remainder of the commission if the customer leaves. You can also offer an even larger commission if they'll take stock over cash.

Once commissions are vested on a schedule that prioritizes customer relationships, a lot of the ugliness that usually defines sales cultures disappears. Salespeople do a better job qualifying customers, the hypercompetition eases up, the backslapping fades, the teams align their expectations and their goals.

It just works better. For everyone.

The old commission model is an anachronism. It's outdated and rewards all the worst behavior. But it is useful for one thing: weeding out the assholes.

There are many incredible salespeople out there who will raise an eyebrow at the idea of vested commissions, then lean in and ask to hear more. Others will sneer and roll their eyes and tell you you'll never be able to hire anyone. They won't listen when you explain and will strut out the door, confident they know better and that you're completely crazy.

Don't hire those people.

Find people who are intrigued by the idea of vested commissions. Find people who realize they can actually make more money this way. Find people who are good human beings in addition to being good at sales. Find people who will care about your mission and be thrilled with the vital role they'll play in making it a reality.

It might not be easy. Especially if there's a ton of competition for talent. There are situations and industries where building a whole new sales culture and organization just isn't feasible. In that case, you just need one. Find a sales leader who understands and values customer relationships—someone who won't stand for egoism or cutthroat competition and who won't hire assholes or mercenaries. That leader will shape the culture of their organization to be more relationship-oriented, until the world catches up to what you're doing and you can implement vested commissions.

These people exist. They're tired of transactional cultures, too. They want to do right by their customers. They want to feel like part of a real team. Hire them.

Chapter 5.7

LAWYER UP

Your company will typically need all kinds of lawyers: for contracts, to defend you from lawsuits, and generally to keep you from making stupid mistakes or falling into traps you never saw coming. Early on you can make do with an outside law firm, but eventually that will get too expensive (you'll frankly be amazed how expensive) and you'll probably need to hire lawyers in-house.

But remember that if you're running a business, every decision involving legal matters is a business-driven decision. Purely legal-driven decisions only happen in court. Your legal team is there to inform your choices, not make them for you. So a "no" from legal isn't the end of the conversation—it's the beginning. A great lawyer will help you identify roadblocks, then move around them and find solutions.

••

Most lawyers excel at two things: saying "no" (or "maybe") and billing you.

That's not necessarily because they're bad lawyers; it's just how the system is set up.

Law firms are generally all about billable hours. The first fifteen minutes they talk to you might be free, but they'll charge for every fifteen minutes after that, or even every five. They'll bill you for the time they spent thinking about your company in the shower. They'll bill you for copies, travel, and postage (with an added handling fee, too). They'll charge extra every time they need to call in someone with a specific legal expertise—so if your lawyer brings you into a conference call with another lawyer, expect a jaw-dropping bill.

I once had a lawyer who started every conversation with a little chitchat—how's the family? Some weather we're having! I didn't want to be rude, so I spent a few minutes chatting. But that very polite, normal small talk meant that questions that should have been answered in 15 minutes or less were going 30 or even 45. And this lawyer cost $800–$1,000 an hour. I was paying him hundreds of dollars to talk about my kid's music recital. After three or four conversations, I realized what he was doing and fired him. I can only imagine how he was inflating his billable hours when I wasn't on the phone.

When you're hiring an outside law firm, you want a lawyer who talks fast and does not care about your children—at least not when they're on the clock.

The good news is that some law firms are shifting to a new model—fixed-price contracts or not-to-exceed contracts where everyone agrees on the price up front. Some legal firms will help with rote company formation and boilerplate legal stuff for a small fee or some stock. And there's a new movement to "open source" many important legal documents—to make generic versions that can work for most businesses.

But even if you use open-source legal docs, you still need a lawyer to handle the details. And that lawyer will probably still bill you from the shower.

So to get the most out of your lawyer, you need to understand how they operate and how they approach their work. Lawyers are trained to think from the competitor's viewpoint or the government's viewpoint or that of pissed-off customers or irate partners or suppliers or employees or investors. Then they look at what you're working on and say, "Doing it this way will almost certainly get you in trouble." Or, on a really good day, "Doing it this way may turn into a lawsuit but we'll probably be able to handle it."

You will never get a pure, unadulterated "yes, go ahead—there's no danger ahead" because there's no ironclad way to prevent a lawsuit. Anyone can sue you for anything—at least in the United States. Customers will sue you for changing something they liked. Competitors will sue you as a business tactic to shut you down. Merit may have nothing to do with it—they'll hammer you with nuisance lawsuits just to drain your coffers and your will.

If you're moderately successful at something disruptive, you'll probably be a target. If you're really successful, you definitely will be.

So the possibility of a lawsuit should always be a risk that you're weighing. But a lawsuit isn't the end of the world. And a "maybe" or even a "no" from your lawyers isn't always a reason to immediately stop what you're doing. You have to weigh their input against the needs of your business and against the fact that you need to take risks to innovate and succeed. That doesn't mean you shouldn't follow legal advice—it just means that legal shouldn't be your only consideration.

Of course, this doesn't apply to anything actually illegal. Or lies. Or any of the basic stuff you need a lawyer for—contracts or HR or the terms you put on your app for protections and privacy. For that sort of thing, do not screw around. Listen to your lawyer and follow their advice explicitly. If you don't have a lawyer on staff, then hire a firm and deal with the bill. You do not want your business to crumble because of a stupid mistake—because you screwed up your employment agreements or your terms and conditions.

But for the gray areas, for the tricky stuff, for the million nuanced opinion-driven decisions that will determine the direction of your company, always remember that lawyers live in a black-and-white world. Legal versus illegal. Defendable versus undefendable. Their job is to tell you the law and explain the risks.

Your job is to make the decision.

The first time I had to deal with a lawsuit was at Apple. I remember feeling like a deer in headlights. Creative, the maker of the second most popular music player after the iPod, sued us over the iTunes interface to transfer songs to the iPod and the technology that enabled it. It wasn't clear-cut whether we had infringed or not, whether we would win or not, and Steve was worried. We'd built Apple their first great new product in years, and now they were getting sued for it.

Chip Lutton, who led all intellectual property legal work at Apple, partnered with me and Jeff Robbin, the VP of iTunes, to find solutions to work around the issues. We proposed different ways to change the product, but in the end Steve ended up making the business decision to settle. In fact he settled for $100 million, tens of millions more than Creative had asked for. He wanted them completely out of our hair, never to return.

It was an interesting lesson in what winning meant. It wasn't a proper legal victory—we never defended ourselves, never went to court—but it was a win for Steve. It was more important for him to never spend another second of his life worrying about this lawsuit than to save money or face.

Pretty quickly after we launched the Nest Learning Thermostat, we got sued by Honeywell. This was a very different lawsuit. They were fully committed to litigating us out of existence. Their strategy was to squash the little guy and all but steal their technology for a pittance. Our legal team was confident we could win—the lawsuit was ridiculous, frivolous, a well-known tactic to slow down a fast-growing competitor. But I'd learned from my experience at

Apple—I couldn't just hand the decision for what to do over to legal.

Lawyers love to win—they will never give up the fight, will battle to the death. But this is business. Death is not an acceptable option. You don't get a great ROI with death.

When you're in any kind of negotiation that includes legal, you always need to work out the fundamental deal points first, before the lawyers get called in—how much you're paying for something, how much you're willing to spend, how long a contract should last for, exclusivity, etc. Get the term sheet roughly approved, then let the lawyers argue the legalese. Otherwise negotiations can drag on forever, with you footing the bill as your lawyers fight with their lawyers.

Nobody wants to deal with that.

That's why even when we were right on the verge of triumph in our lawsuit with Honeywell, we settled out of court. By that time Google had bought our company and Honeywell was a major customer of theirs. It didn't matter that we were right and Honeywell was wrong; it was a business decision. Google decided that paying off Honeywell and maintaining that relationship was preferable to going to court—especially since the cost of the settlement came out of Nest's coffers and not Google's.

It pissed us off no end. We would have won, but instead Nest got stuck with the bill. It was enraging. But it was the right call for Google.

The best lawyers understand that. They don't only think like a lawyer. They take into account all their training and knowledge, but also weigh business objectives. They can help you understand the risks while also being very aware of the benefits.

They give you well-reasoned advice rather than tell you what you can and can't do. They realize their voice is part of a chorus. And as you work together and get a feel for each other and they get a handle on what the competitive landscape looks like and who your customers and partners are, great lawyers will loosen up a little. It takes most lawyers months or even years working with a company to really understand which risks are worth worrying about and which can be mostly ignored. But a well-seasoned, experienced, and business-practical lawyer who can communicate risk effectively can be worth their weight in gold.

Getting a lawyer like this generally requires you to hire one in-house. Usually you'll know it's time to start the hiring process when legal gets way too expensive—too many hours to deal with the same agreements and questions, too much back-and-forth, too many times you need them to search for too-rare specialists.

An in-house lawyer won't solve your need to hire specialists—for taxes, HR, fund-raising, M/A, IP and patents, government rules—but they will help you negotiate the bills after you get them. Because there's always room for negotiation—especially with lawyers. An experienced lawyer who understands a law firm's business model and knows the tricks can look at their bill and ask why it took this many hours for this task or why this conversation was billed in this way.

When considering your first legal hire, you may be tempted to hire a generalist—someone who can do a bit of everything. People think that will cut down on their need to hire outside specialists. But it's the opposite.

At this moment, you're not hiring for breadth. You need to understand what's at the core of your company—what your business is ultimately about—and hire for those specific legal specialties.

Too many times I see companies where intellectual property is their biggest differentiator, but they get a regular contracts lawyer to run the legal team. It's a costly mistake. The lawyer ends up outsourcing all IP legal work, negating any cost savings, and then can't even provide guidance to outside counsel. When the first legal hire doesn't have experience and expertise in critical areas, the legal team ends up weaker for it—more risk averse, less flexible, less able to work with the rest of the business to creatively solve problems and build effective long-term legal strategies for the company.

At Nest, we knew from the start that everything would come down to IP. Nest's special sauce was always going to be our technological innovations. And those innovations would have to be aggressively patented to keep them out of the hands of the competition.

So our first lawyer was Chip Lutton, the same lawyer I'd worked with on the iPod lawsuit.

We needed a leader who already had a deep understanding of the issues that would be at the heart of our business, who could think that way from day one and build his team with that perspective in mind. And we needed someone who could act as a moral compass, who could go toe-to-toe with execs and engineers and marketers.

We needed a leader who could lead.

Who would be respected and thoughtful enough to be an active part of product development.

Chip and his team were never back-office. They were always in the thick of it with us, thinking through product features, making sure we could defend our patents, looking over our marketing copy, beating back lawsuits. And getting into fights with me.

Like the one about the baby.

In June 2015 we launched the Nest Cam, a video camera that you could use for security or as a pet cam or baby monitor. Also, in the United States, any electronics intended to be used in a baby's room must come with this warning:

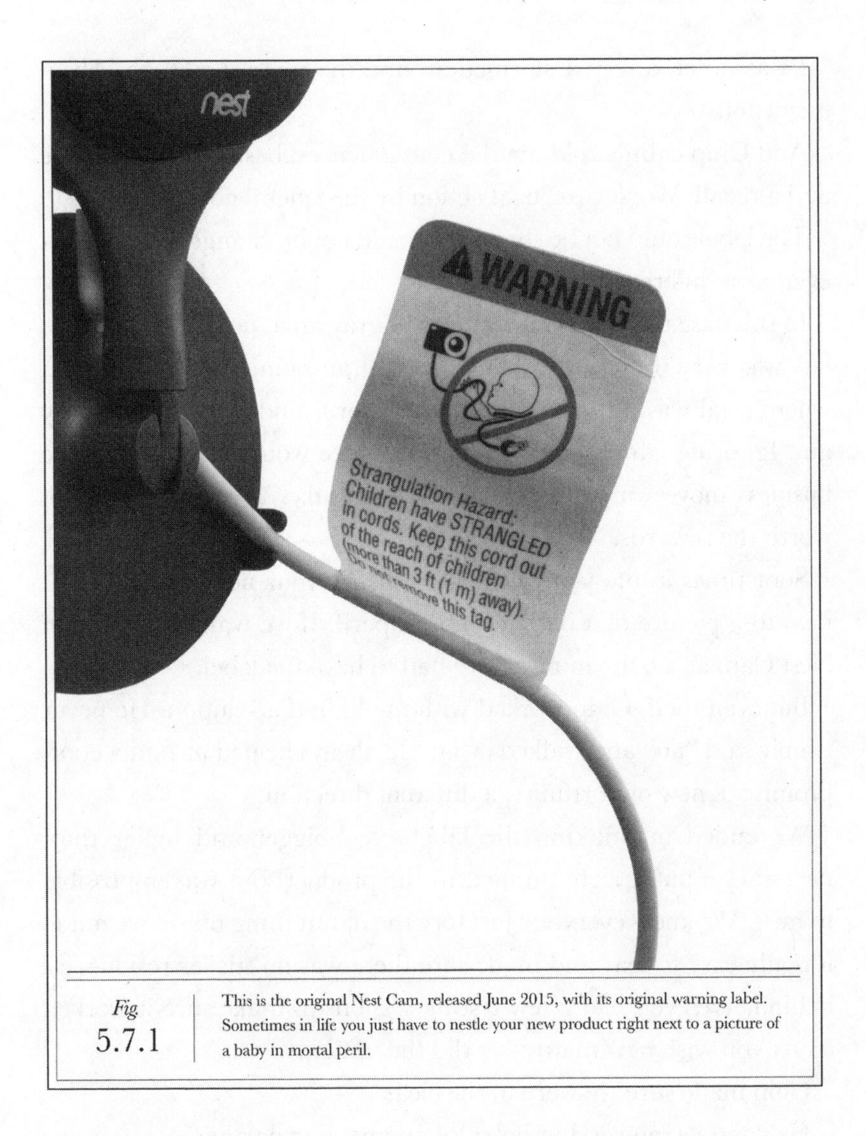

Fig 5.7.1 | This is the original Nest Cam, released June 2015, with its original warning label. Sometimes in life you just have to nestle your new product right next to a picture of a baby in mortal peril.

And I said, "No way. We are not launching our new product with a picture of a strangled baby."

We'd already put the strangulation warnings all the way through the app, in the installation instructions, the manual, the setup. We made it impossible to ignore. None of our competitors' products had gone to the depth we had. And they all had the same cables!

I was upset, angry. I stormed around the room. I said no. Absolutely not.

And Chip calmly told me the consequences: best case a hefty fine and a recall. Worst case legal action by the federal government.

The label could not be shrunk. It could not be changed. It couldn't even be a different color.

In this case there was no nuance, no gray area, no discussion. The law was very prescriptive. This was not an opinion-based decision where legal was just one voice in the chorus and I could follow my gut. Ignoring Chip's advice in this instance wouldn't be a strategic business move—it would be a stupid mistake. The risks were not worth the rewards.

Sometimes in life you just have to nestle your new product right next to a picture of a baby in mortal peril. If we wanted to market Nest Cam as a baby monitor, we had to have the label.

But even then, Chip worked with me to find a solution. He never simply said "no" and walked away. He always helped us find a compromise, a new opportunity, a different direction.

We ended up making the label even bigger and uglier than necessary—put it right up next to the product so it was impossible to miss. We knew everyone just tore the damn thing off, so we made it really easy to tear and made sure there was no sticker residue left behind. We even had a few testing sessions to make sure it worked (don't you wish new mattresses did the same?).

Chip made sure we were in the clear.

He's a great lawyer, but he's also an amazing partner.

That's what you're ultimately looking for. You don't want a lawyer who thinks their only job is to point out the sinkhole you could fall into, to block your way. Hire someone who will help you find a new path. Hire someone who will build a bridge. Hire a lawyer who doesn't just think like a lawyer.

Part

VI

BE CEO

The connected home was going to really blossom with the Nest Protect smoke and CO alarm.

Nest Protect would act as a temperature and humidity sensor for the Nest Learning Thermostat so you could have climate control room by room. It would use its motion sensor to detect when nobody was home so the thermostat could turn off the heat or AC right away to save energy. And it would use its voice to do more than just warn you of danger—the vision for Nest Protect was for it to be a really great speaker. There's a smoke alarm in every room in the house, so we planned to have it play music and even work as an intercom. You'd call up to the Nest Protect in the kitchen that dinner was ready and the smoke alarm in your kid's bedroom would carry your voice.

Now add a video camera or a smart lock into the mix and you've got a built-in security system with sensors in every room and alarms throughout the house. With every new Nest you'd install, your old Nest products would get better. Do more things. Open new avenues of convenience and possibility. And it would require very little from you. It would all just . . . work.

The whole point of the connected home was for it to be effortless. For your house to take care of you rather than you taking care of it.

And we weren't blind to the dozens of other connected products sprouting up around us once the Nest Thermostat showed what was possible. Instead of treating them as competition to be squelched, we nurtured the ecosystem with a low-powered networking technology called Thread. If you built a decent smart device, you could hook into the Nest system and work with Nest products—your smart ceiling fan could connect to our thermostat or Nest could tell your smart lights that you were on vacation so they could make it look like you were home and fool potential thieves.

Nest was building a platform—an ecosystem of our own and third-party products all controlled by one application—that would elevate the connected home into something truly magical. We were going to weave a tapestry of technologies that would fundamentally change what home could be.

That was the vision, anyway.

The vision that Google bought for $3.2 billion in 2014.

Google had been close to Nest from the beginning. I'd shown Sergey Brin some prototypes before we launched and Google wanted to buy our company back in 2012. They wanted to help us reach our vision faster. When we declined, they offered to invest instead.

In 2013, as we were in the midst of another successful funding round, they pushed to acquire us again.

I knew that if they were that eager to buy Nest, it meant they could be getting serious about building smart-home hardware. And if Google was getting serious, then Apple, Microsoft, Amazon, Facebook, and the other tech behemoths could be looking into it, too. Nest had started this snowball rolling and now it was causing an avalanche.

If we weren't careful we could quickly get buried.

Nest was doing really well, selling our products as fast as we could make them. We could have just kept building the thermostat—its impact had already eclipsed our wildest expectations. People were putting thermostats—thermostats!—under their Christmas trees. David Letterman and Kanye West wrote to us asking for our product when we were sold out.

But we were dead set on building a platform—a large and meaningful one that could last decades—and that would require vast resources. Huge companies like Google or Apple with other highly profitable revenue streams and tons of products could quickly displace us with their own platform. All they'd need to do was to announce a plan to enter the connected-home space. It wouldn't matter if their platform was any good—when a giant company makes an announcement, that alone puts a thumb on the scale. They could lure away all our potential partners and developers or just keep them from fully committing to us while they took a "wait and see" approach.

I had seen too many successful products and platforms from little startups die when bigger players moved in and sucked up all the oxygen in the room.

But by joining Google we wouldn't just be protecting ourselves—we'd be accelerating our mission. That's what really excited the exec team. The potential for growth.

So after a lot of deliberation and nervousness, we decided, as a team, that it was the right time to sell. We had a strong hand—plenty of funding, more investors circling, and solid unit economics. Our business was lean, our thermostats profitable, another product was shipping, and we had more in development.

And Google had made a commitment to invest $4 billion in our connected home platform over five years and provide necessary resources—servers, AI algorithms, developer relations. We agreed that they'd set aside what they'd started building in connected-home hardware and focus entirely on Nest. We agreed to coordination meetings every other week with the teams whose technology we'd need to integrate with.

We had plenty of concerns about a culture clash, but the Google team reassured us that Nest's mission-driven culture would set a new standard and help drive a cultural evolution at Google. They told us we'd get a huge boost in sales and that we'd be able to bring our platform to life years faster than we could as an independent company.

They told us this was going to be a beautiful marriage.

After three or four months of intense discussions, we both walked into the wedding in January 2014 believing we were going to stay together for good. That this was going to work. That we would make it work. Both companies had the very best intentions.

But everyone knows what the road to hell is paved with.

Within hours of the acquisition, Nest tried to calm a wave of bad press by publicly declaring our culture and systems completely separate from Google. After that, there was organ rejection. The natural antibodies at Google detected something new, different, foreign and did everything they possibly could to avoid or ignore it. They would smile and smile but the promised meetings, the oversight from Google management, the plans we'd made to integrate—they all started to fall apart.

Even the basic, no-brainer stuff stalled out: Can we sell Nest products in the Google Store? No. Not for at least a year. Can we get off Amazon Web Services and onto Google Cloud? No, not without a huge set of changes. And it'll actually be more expensive.

That shouldn't have been a shock, though. Everything was more expensive.

In 2014, just before the Google acquisition, Nest spent around $250,000 per employee per year. That included decent office space, good health insurance, the occasional free lunch, and fun perks from time to time.

After we were acquired, that number shot up to $475,000 per person. Some of the increase was due to corporate red tape and increased salaries and benefits, but a lot of it was the added perks of free buses, free breakfast, lunch and dinner, tons of junk food, gleaming conference rooms with full A/V setups, and new office buildings. Even IT was expensive. It cost $10,000 per year to connect each employee's computer to the Google Network and that didn't even include the price of the laptop.

Of course, Nest wasn't perfect, either. We had so many different projects going at once that we had indigestion. The Nest Secure security system was perpetually delayed as we launched a second-generation Nest Protect and a third-generation thermostat. We'd also bought a company called Dropcam and created the first Nest Cam and added it to the Nest app and spent countless hours trying to integrate with Google and figure out email addresses, corporate security, whose servers had what data on them, privacy policies, etc.

And despite being part of Google, we made little attempt to be Googley, to truly join the culture. A small contingent of Nesters had come from Apple, where Google was Enemy #1 and they had to be talked down from a cliff. But most of us just liked our way of doing things. We didn't want to be Googley. I wasn't about to wear a propeller baseball hat, like new Googlers do. I can understand why we stood out like a sore thumb, why we weren't welcomed with open arms.

But even with all that, the acquisition wasn't a complete disaster. It was a work in progress.

Our brand and unit economics were strong. We were still growing quickly. In fact the Google acquisition had given a bunch of retailers the confidence to start selling Nest products in their stores. Developers for our ecosystem were also signing on in much larger numbers. And we were making minor inroads with certain teams at Google, though not nearly the progress we'd expected. But we had time. The plan was five years of runway to build a true smart-home platform. And there were so many amazing people at Google, so many incredible teams building incredible technology that we could work with to create something truly spectacular and important. We just had to keep pushing. We could get there.

Then, in August 2015—a little over a year after the acquisition closed—Google cofounder Larry Page called me into his office. He said, "We've got an exciting new corporate strategy for the company. It's called Alphabet. And we want Nest to be the model for how to do it right."

They were restructuring Google—creating an umbrella company called Alphabet to hold Google and all its "other bets" as subsidiaries. That way Wall Street could see the health of Google's Search and Ads business without anything clouding the financial picture. The "other bets"—like Google Fiber, Calico, Verily, Capital G, Google Ventures, Google X with its myriad of "moonshot" projects, and, of course, Nest—would become independent sister companies that were no longer part of Google. Suddenly Nest would become one of the biggest, most public, and most expensive of the sisters.

For sixteen months we'd been focused on trying to integrate with Google, to become one with the mothership and get all the goodness we needed to accelerate our vision. That integration and technology access was the main reason we agreed to be acquired in the first place. But Larry told me that was over. New direction. New strategy.

"How long have you been thinking about this?" I asked.

"Years," he said.

"How many people at Google have been thinking about this?"

"Three or four for a couple months. You're one of the first people I'm telling."

I thought, "Great! Thanks!" But I said, "Okay—we'll need to understand the details to make sure we're all aligned. How much time do we have to dig in and work out a concrete plan? A few months?"

I knew better than to say "no" without a detailed rebuttal. But I needed to buy time to figure out how to undo this—to get our team a better deal.

"We don't have a few months."

"Eight weeks?"

"No."

"A month?"

"We're going to announce it next week," he said. "We're a public company—it would be a disaster if this leaked to the press. This is just a finance and accounting change; don't worry about it, we'll figure it out."

I was shocked. Speechless. So much for delayed intuition. I thought to myself, Fire, ready, aim.

There was no plan in place. None. I'm all for "Do. Learn. Fail," but you can't turn an entire company upside down without at least a semblance of a strategy. What should have been a data-driven decision had turned into an opinion-driven one.

Larry told me he'd been watching how Warren Buffett did it at his firm. He'd even flown to Nebraska to talk to him about it. Berkshire Hathaway buys unrelated companies that are run separately and it works great. "Why can't we do the same?"

I pointed out that Berkshire Hathaway buys companies that are ten, fifteen, fifty years old. They're fully formed, have plenty of revenue. They're grown, healthy adults. Alphabet's other bets were infants, toddlers, adolescents trying to find themselves. They were still scrambling to innovate, trying to find a path to profitability. The fundamentals were totally different.

But that didn't matter. The steamroller was already on its way.

Within twenty-four hours of the Alphabet announcement, Google Facilities said, "You're not part of Google anymore, so you'll need this" and handed us a bill in the millions for our new office that they'd just remodeled.

To add insult to injury, our cost per employee more than doubled—a 2.5x increase. And nothing was different—everyone at Nest was doing the same job in the same place. Now we just had to pay for it ourselves, including an Alphabet corporate tax for any service that Google supplied. So the basics we relied on—IT, legal, finance, HR—got instantly more expensive. Sometimes ridiculously expensive. We heard, "Sorry we have to do this. It's a FASB [Financial Accounting Standards Board] requirement. No way to get around it since we're a public company."

In that same instant, the technology teams at Google that we'd finally started integrating into happily washed their hands of us. "You aren't Google," they told us. The antibodies were on full-scale attack.

It was shocking to watch the speed with which their priorities changed.

But the worst part was the bullshitting.

I started getting allergic to the word "thoughtful." Every time Google senior management wanted us to swallow some new strategy, they'd tell us it was, despite appearances, extremely thoughtful. "We're being thoughtful about the Nest integration into Google." "We're going to be thoughtful about your transition to Alphabet." "Matt and Tony, don't worry—we've been thoughtful about this."

I'd hear it and think, Oooohhh noooo, here it comes again, as they thoughtfully bullshitted me, then expected me to pass that bullshit on to my team.

Matt and I had been beating the drum for integration for over a year, but now we had to turn 180 degrees and beat the drum for Alphabet. I had to tell our team that things were going smoothly even as I watched Google leadership "thoughtfully" jury-rig a plan as the transition was already under way, then change that plan constantly over the next few months. It was a complete mess of Alphabet implementation meetings each week—finance, legal, IT, sales, marketing, PR, facilities, HR. One day we'd be told that this is how they'll bill for buses or facilities or legal service, and then two weeks later it would get rethought.

Then, as the cost of Nest being Alphabetized became known, the new finance regime swooped in.

The Alphabet steering committee said we need to rationalize Nest's expenses and get to profitability faster. They pointed out that we weren't hitting our sales numbers. I pointed out that they'd made those numbers up. They'd assumed that Nest products would be sold in the Google Store and that would boost our sales by 30–50 percent. But then the Google Store had stalled and waffled and our sales actually fell as customers wary of Google's privacy policy stayed away.

When it was clear I wasn't going to budge, Larry told me that we had to reach profitability. "I need you to get bold and creative and figure out how to cut everything by fifty percent." And he literally meant everything: head count, expenses, and our road map.

"WHAT?!" I said. Nothing had changed—our agreement was the same, our plan was the same. But now they wanted me to lay off half our team, most of who we had just hired over the previous months.

Larry told me not to worry about firing everyone. He said there were plenty of open jobs at Google that they could easily move into. And I thought, Has this guy ever had to personally lay off people in his life? You can't just play with people's lives like that.

But Google wanted to show Wall Street that one of their businesses beyond search and advertising could actually be profitable. Every other piece of hardware they made—the cell phones, the Chromebooks—was bleeding money. Nest was the only one that had a shot of reaching profitability, so they focused all their attention on us.

But there was no way—absolutely no way—I was going to hack Nest in half.

We proposed 10–15 percent reductions but flatly refused to change our road map. We would not back down from our mission.

Needless to say, tensions were high.

Four months later, I got hit with another bombshell.

Larry Page told me he wanted a divorce.

He was going to sell Nest.

However, Larry didn't actually say it. Bill Campbell, my mentor and a mentor to Larry, asked me to stay late one day after a board meeting, just days before the end-of-year holidays. After everyone cleared out, he and Larry were left. Bill took one look at me and said, "I'm just going to cut to the chase. Larry will have a hard time saying it and I'm not going to bullshit you. Larry wants to sell Nest. I don't understand it, but this is what he wants to do."

Larry looked at Bill, shocked. "Well, you didn't have to say it like that!"

But he did. Bill knew me. And Larry didn't. Not really. I suspect he wanted Bill in the room because he was worried I'd freak out. He wanted a witness and a buffer for our breakup in case things got heated.

Instead I sat there in silence, trying to hear every word, watching every micro-expression on their faces as Larry tried to explain the situation.

Then I said, "Larry, you bought Nest. You can sell Nest if you want. But I'm absolutely not going with it."

I was done.

Bill looked at Larry and said, "I knew it. I told you that would be his answer."

Even now, I'm not exactly sure why Google decided to sell. Maybe it all came down to the culture clash—maybe Larry thought we were too far apart, too incompatible. When I asked, they gave me the usual reasons: we've decided Nest is no longer strategic, it's costing us too much money. But even as Google had changed, our agreement hadn't—we'd been up front about our road map, our plan for the future. They knew when they signed the deal that we were an expensive investment. And they'd been more than willing, in fact had been eager, to fund our vision less than two short years ago.

"We have a new finance strategy now," they said. And that was that.

Bill was beside himself after the meeting. "You guys have popular products, solid economics and growth, and a pipeline of new products with real potential. You have way more going for you than most of the projects around the company," he told me, his head in his hands. "It doesn't make any sense whatsoever. We were just getting started!"

Over his objections, Google brought in the bankers so I could help them "preserve asset value." And since I had already said I was leaving, the only thing I could do was try to minimize the damage, to smooth the transition for my team as much as I could. I fell into the role of being a good soldier. I did what I was told, helped the bankers prepare the documentation for a sale. They began shopping Nest around in February 2016.

The bankers worked down their possible-acquirers list and a few companies came to the table. At the head of the line was Amazon.

The bankers asked Larry if he'd sell to Amazon and he said, "Yeah, I think I would." I was shocked again—sell to one of their competitors? It felt like another slap in the face.

As talks moved forward, I stuck to my word. I left Nest. Walked away from our marriage. They said they wanted a divorce so I gave them a divorce.

And then, months after I left, Google changed their mind. Again.

They decided not to sell Nest after all.

In fact, they decided that Nest was better as part of Google rather than being lumped in with the "other bets" of Alphabet. And Nest was reabsorbed.

It was the strategy du jour—join Google, leave Google, join Google. All the while, Nest's management team had to stand in front of employees and promise that everything was going to be great! But there was no denying that the flip-flopping was painful—for our customers, for the team, for their families. Executive management seemed to have an utter disregard for our employees and the work they were trying to achieve.

In the end, when Google reabsorbed Nest in 2018, they moved ahead with the 10–15 percent reductions I'd proposed back in late 2015. Rejoining the mother ship also eliminated the overhead costs of Alphabet—that additional $150,000 per head and the myriad taxes and higher fees. Suddenly it seemed Nest was a great investment again.

I can't explain it. Just as I never learned their true reasons for selling Nest, I never heard an explanation for why they decided to keep it. Maybe the fact that Amazon was interested made Larry realize that Nest was a valuable asset after all. Maybe it was all an elaborate game of chicken to get me to toe the line and cut costs. Maybe they never had a real plan to begin with and this all happened because of some exec's casual whim. You'd be surprised how often that's the reason behind major changes.

People have this vision of what it's like to be an executive or CEO or leader of a huge business unit. They assume everyone at that level has enough experience and savvy to at least appear to know what they're doing. They assume there's thoughtfulness and strategy and long-term thinking and reasonable deals sealed with firm handshakes.

But some days, it's high school. Some days, it's kindergarten.

That was true when I first joined the C-suite at Philips and when I became a VP at Apple and when I was CEO of Nest and when I entered the ranks of Google execs. All these jobs felt incredibly different, but at their core the responsibilities were the same—it was less and less about what you were making and more about who you were making it with.

As CEO, you spend almost all your time on people problems and communication. You're trying to navigate a tangled web of professional relationships and intrigues, listen to but also ignore your board, maintain your company culture, buy companies or sell your own, keep people's respect while continually pushing yourself and the team to build something great even though you barely have time to think about what you're building anymore.

It's an extremely weird job.

So if you've climbed to the top of the corporate mountain and now are frostbitten and oxygen-deprived and wondering when the Sherpa will get there, then here are some of the things I've learned.

Chapter

6.1

BECOMING CEO

There's nothing exactly like being a CEO, and nothing to prepare you for it—not even being the head of a huge team or division of a company, firmly in the C-suite. In those positions, there's always someone above you—but the buck truly stops with the CEO. And as CEO, you set the tone for the company. Even if there's a board, partners, investors, employees—ultimately everyone looks to you.

The things you pay attention to and care about become the priorities for the company. The best CEOs push the team to strive for greatness, then take care of them to make sure they can achieve it. The worst CEOs care only about maintaining the status quo.

There are generally three kinds of CEOs:

1. **Babysitter CEOs** are stewards of the company and are focused on keeping it safe and predictable. They generally oversee the growth of existing products that they inherited and don't take risks that might scare executives or shareholders. This invariably leads to the stagnation and deterioration of companies. Most public company CEOs are babysitters.

2. **Parent CEOs** push the company to grow and evolve. They take big risks for larger rewards. Innovative founders—like Elon Musk and Jeff Bezos—are always parent CEOs. But it's also possible to be a parent CEO even if you didn't start the business yourself—like Jamie Dimon at JPMorgan Chase or Satya Nadella at Microsoft. Pat

Gelsinger, who recently took over the Intel CEO position, seems to be Intel's first parent CEO since Andy Grove.

3. **Incompetent CEOs** are usually either simply inexperienced or founders who are ill-suited to lead a company after it reaches a certain size. They are not up to the task of being either a babysitter or a parent, so the company suffers.

••

The job is to give a shit. To care. About everything.

I remember going to the Aston Martin factory once to have a meeting with the CEO. It was 9 a.m. and pouring as we drove through the lot. At one point we had to stop the car as a guy covered in bright yellow raingear and galoshes hurried across our path. When we got to the meeting, in comes the guy in the galoshes. It was the CEO. Andy Palmer had been walking the lot, personally inspecting each car that came off the line.

The CEO sets the tone for the company—every team looks to the CEO and the exec team to see what's most critical, what they need to pay attention to. So Andy showed them. He stepped into the rain and looked at the engines, the upholstery, the dashboards, the exhaust pipes, everything. He rejected any car that wasn't perfect.

If a leader gets distracted from the customer—if business goals and spreadsheets full of numbers for shareholders become a higher priority than customer goals—the whole organization can easily forget what's most important.

So Andy showed every person in that company where their priorities should be. He didn't care how much it cost to reach perfection, how many times a car had to get retooled, reworked. The important thing was to deliver exactly what the customer expected. To over-deliver.

If you want to build a great company, you should expect excellence from every part of it. The output of every team can make or break the customer experience, so they should all be a priority. [See also: Chapter 3.1: Making the Intangible Tangible.]

There can't be any functions that you dismiss as secondary—where you casually accept mediocrity because it doesn't *really* matter.

Everything matters.

And it's not just about you.

If your expectations are that everyone puts out their best work, if you're looking at customer support articles that will be posted on your website with the same critical eye as engineering or design, then the technical writers of those articles will feel the pressure, will bitch and moan, will get stressed out, and then they will write the most incredible support articles of their lives.

That's not a hypothetical example. I read most of the key customer support articles for all our products at Nest. Those articles were the first thing a customer having issues would see. This customer would be frustrated, irritated, right on the brink of anger. But a spectacular experience with support could instantly turn that frustration into a moment of delight—into a customer who would be with us forever. I couldn't ignore the importance of that moment because it was "just" support. So I read the articles. And I critiqued them. In fact, through that process I learned things about the product experience that I didn't know—and didn't like—and worked to fix them.

And I read these articles with the support and engineering teams around me—I wanted all of us to question the content, to make sure our support site was as crisp and easy to understand as our marketing and sales materials. I showed them through my actions that what they did was important. And when they came back with a new version, I read that, too. And I tore it apart until each article told a story. Until they gently guided customers to understanding rather than just barking confusing instructions at them.

When you truly give a shit, you care, you don't let up until you're satisfied, you pick things apart until they're great.

People will hand you something that they worked on tirelessly for weeks, that they've thought through and are proud of, that's 90 percent amazing. And you will tell them to go back and make it better. Your team will be shocked, stunned, possibly even dejected. They'll say it's already so good, we've worked so hard.

You'll say good enough is not good enough. So they'll march out the door and do it again. And, if necessary, again. They might get so tangled up that it'll be simpler to just start from scratch. But with each iteration, each new version, each regroup and reimagining, they'll discover something new. Something great. Something better.

Most people are happy with 90 percent good. Most leaders will take pity on their teams and just let it slide. But going from 90 to 95 percent is halfway to perfect. Getting the last part of the journey right is the only way to reach your destination.

So you push. Yourself. The team. You push people to discover how great they can be. You push until they start pushing back. In these moments, always err on the side of almost-too-much. Keep pushing until you find out if what you're asking for is actually impossible or just a whole lot of work. Get to the point of pain so you start to see when the pain is becoming real. That's when you back down and find a new middle ground.

It's not easy. But all that attention, that care, the quest for perfection—they'll raise the team's own standards. What they expect of themselves. After a while, they'll work incredibly hard not just to make you happy, but because they know how much pride they feel when they do world-class work. The entire culture will evolve to expect excellence from each other.

So your job is to care.

Because you're it. You're the top of the pyramid. Your focus, your passion, trickles down. If you don't give a shit about marketing, you'll get shitty marketing. If you don't care about design, you'll get designers who don't care, either.

So don't worry about picking your battles. Don't rack your brain trying to decide which parts of your company need your attention and which don't. They all do. You can prioritize, but nothing ever comes off the list. Avoiding or ignoring any part of your company only comes back to haunt you sooner or later.

At Nest I met with product teams and marketing every two weeks and customer support every month, and I'd have a meeting with every team at the company at least twice a year. Even if you were building internal software tools for HR or operations, eventually you'd be called to show your strategy. I'd watch the presentation and then dig in—Do we have the right IT back end to do this? How are you planning to get around this issue? How can other people on the team help? How can I help?

It didn't matter if this team was building internal tools that customers would never see. The company depended on those tools, and internal customers should be treated as well as external ones.

So I listened closely, gave them my full attention (don't be checking your phone or computer), and helped them move past their roadblocks. That's often all it takes.

And if you're not an expert at internal software tools or PR or analytics or growth or whatever needs you to have an opinion today—if you're not sure what's great and what's just okay—then ask questions. I love asking dumb, obvious questions or questions from the customer's perspective—usually three or four "Why is that . . . ?" "Why did that . . . ?" questions will get to the root of what you're trying to understand, then you can dig in further. And if that's not enough, then call in the experts. Bring in experienced people from your team (or sometimes from outside it) who can confirm your inklings or steer you in the right direction until you learn enough to trust your gut.

You don't have to be an expert in everything. You just have to care about it.

No matter your leadership style, no matter what kind of person you are—if you want to be a great leader, you have to follow that one cardinal rule.

The other commonalities of successful leaders are just as straightforward:

- They hold people (and themselves) accountable and drive for results.
- They're hands-on, but to a point. They know when to back off and delegate.
- They can keep an eye on the long-term vision while still being eyeball-deep in details.
- They're constantly learning, always interested in new opportunities, new technologies, new trends, new people. And they do it because they're engaged and curious, not because those things may end up making them money.
- If they screw up, they admit to it and own their mistakes.
- They're not afraid to make hard decisions, even when they know people will be upset and angry.
- They (mostly) know themselves. They have a clear view of both their strengths and challenges.
- They can tell the difference between an opinion- and data-driven decision and act accordingly. [See also: Chapter 2.2: Data Versus Opinion.]
- They realize that nothing should be theirs, even if the genesis was with them. It all has to be the team's. The company's. They know their job is to jubilantly celebrate everyone else's successes, to make sure they get credit for them, and hold little for themselves.
- They listen. To their team, to their customers, to their board, to

their mentors. They pay attention to the opinions and thoughts of the people around them and adjust their views when they get new information from sources they trust.

Great leaders can recognize good ideas even if those ideas didn't come out of their own mouths. They know that good ideas are everywhere. They're in everyone.

Sometimes people forget that. They firmly believe that if they didn't think of it, it's not worth thinking about. That sort of egoism can also extend way beyond individuals—many CEOs get so wrapped up in their own companies that they dismiss the competition. If it wasn't invented here, it can't possibly be any good.

It's the kind of thinking that kills companies, that collapsed Nokia, that toppled Kodak. It's probably what was in Steve Jobs's head when he refused to meet with Andy Rubin.

I'd known Andy, the founder of Android, since we worked together at General Magic. And in the spring of 2005 he heard through the rumor mill that Apple was working on a phone. So he called me up. He wondered if Apple might be interested in investing in or possibly buying Android, his latest project to create an open-source phone software stack.

I went straight to Steve. I pointed out that it was a capable team and a great technology. We could use their tech to get a jump-start on the iPhone and eliminate a potentially formidable future competitor in one acquisition.

In typical Steve Jobs fashion he said, "Fuck that. We're doing it ourselves. We don't need any help."

Part of Steve's reaction was clearly motivated by confidentiality—the other part was Not Invented Here Syndrome.

But I knew Andy and the threat his project might pose, so I brought up the topic again two weeks later, in front of Apple executives and iPhone development leads. Steve didn't want to hear it. A week later

I emailed Andy and got no response. A month after that, we saw the announcement that Google had bought Android.

It's hard to imagine what could have happened if Steve had had just one meeting with Andy to understand his strategy, never mind buy his company. How would the world have changed? How would Apple have changed?

It's poison to think great ideas can only come from you. That you alone can hoard them in one place. And it's stupid. Wasteful.

A CEO has to recognize fantastic ideas—regardless of their source. But Apple was Steve's baby and every other baby on Earth was uglier and dumber than his.

I read a study the other day that said that the brain patterns of entrepreneurs thinking about their startups are extremely similar to those of parents thinking about their children. [See also: Reading List: "Why and how do founding entrepreneurs bond with their ventures?"] You are literally a parent to this business—you love it like you birthed it, like it's a part of yourself.

And sometimes your love for your child blinds you to its faults or the brilliance of other ways of doing things, other ways of thinking.

On the other hand, that all-consuming love can help you push your company forward.

As a parent you never stop worrying about your kid, planning for your kid, pushing your kid to do better, be better. A parent's job isn't to be friends with their kids all the time—it's to build them into independent, thoughtful humans who will be ready and able to thrive in the world one day without their parents.

Kids often resent them for it. Cry, slam doors, wail in anguish when you make them turn off the TV, get their homework done, get a job. But you can't be a good parent if you're worried about your kid being mad at you.

Sometimes your kid won't like you.

Sometimes your employees won't, either. Sometimes they'll hate your guts.

I remember walking into meetings and everyone rolling their eyes and sighing. I could see it on their faces, "Oh fuck, here we go again." They knew I was going to keep harping on that one thing that everyone was sick and tired of hearing about. That one thing that was already 90 percent great and that would be so much work to change—too much work—but that I knew in my gut was the right thing for our customers.

It's not a great feeling, having twenty people look at you like that. Like you're being ridiculous, unreasonable. Like what you want is impossible.

It's how we looked at Steve Jobs when he told us five months before the first iPhone shipped that it needed a glass front face covering the display instead of plastic. The front face is the most important part of the hardware—it's the surface you touch constantly.

He realized plastic wouldn't cut it. If we wanted it to be great, it would have to be glass. Even though we had absolutely no idea how to do it. Even though he knew we'd all have to work nonstop until we got it right, sacrificing our time with our families, our plans and vacations.

But Steve was a parent CEO. A pushy parent. A tiger mom. He knew if we kept pushing, together, we'd figure it out. The sacrifices would be worth it.

And he was right. That time. But not every time. Steve took a lot of risks, made bad decisions, launched products that didn't work—the original Apple III, the Motorola ROKR iTunes phone, the Power Mac G4 Cube, the list goes on. But if you aren't failing, you aren't trying hard enough. He learned from the screwups, was constantly improving, and his good ideas, his successes, totally wiped away his failures. He was constantly pushing the company to learn and try new things.

That's how he earned the team's respect. Even when the product took a turn, a huge pile of extra work fell on everyone's shoulders and we knew Steve wouldn't delay the schedule by a millisecond.

It drove us crazy, but the team respected his dedication to getting it right.

In this job, respect is always more important than being liked.

You can't please everyone. Trying can be ruinous.

CEOs have to make incredibly unpopular decisions—lay people off, kill projects, rearrange teams. Often you'll have to take decisive action, hurt people to save the company, to cut out a cancer. You can't skip surgery because you don't want to upset Team Tumor.

Delaying hard decisions, hoping problems will resolve themselves, or keeping pleasant but incompetent people on the team might make you feel better. It may give you the illusion of niceness. But it chips away at the company, bit by bit, and erodes the team's respect for you.

It turns you into a babysitter. And kids may like the babysitter at first—it's nice to go to the local park, to watch movies and eat pizza. It's fun for a while. But eventually kids want to go further, do more. They want to go skateboarding. They want to explore. So they might start testing their boundaries to see what they can get away with. They might roll their eyes when the babysitter tells them what to do. Because a babysitter is not a parent. All kids need someone they respect who truly knows them. Who gives them a push at the right time, who helps them grow.

And they need someone they can project their hopes and aspirations onto.

In the past, in the misty days before you could Google everything about everyone, that's what people did with their leaders. That was one of the things that enabled leaders to succeed. People could believe in, trust, and follow an idealized version of Lincoln and Churchill, Edison and Carnegie.

When your team knows too much about you as a person, not just you as a CEO, they start dissecting your personal life to try to understand your decisions. Your motivations. Your ways of thinking.

That's not only a distracting waste of time, it's counterproductive. When you explain why you're doing something, it should be all about the customers, not about you.

So it's wise to stand alone—not to let anyone at work get too close. Even if you wish you could just grab a drink with your team like you used to.

It's a cliché to say "It's lonely at the top." But it's also true.

Most people assume being CEO is a hard job—stressful, busy, high-pressure. But the stress is one thing; the isolation is another. You might have a cofounder, but you shouldn't have a co-CEO. It's a one-person job and you're all alone up there.

And just because you're in charge does not mean you're in control. You plan out your day, think you're finally going to have some time to talk to people, look at the product, meet with engineering. Then your day disappears. There's always some new crisis, some new people problem, someone quitting, someone complaining, someone falling apart.

And you can never tell if you're getting it right. When you're an independent contributor, you can typically look at something you've made that week and be proud of it. When you're a manager, you can look at the collective achievement of your team and feel a sense of accomplishment and pride. When you're a CEO, you dream that maybe, ten years down the road, some people will think you did a good job. But you can never tell how you're doing in the moment. You can never sit back and look at a job well done.

This job can suck you dry if you let it.

It can also be one of the most liberating experiences of your life.

Since I was a kid, I tried to convince people to follow my crazy ideas. I spent so much time and energy and emotion desperately trying to get them to do things differently. The crazier the idea, the more it cut against the grain, the longer and harder I had to fight for it.

And so often the answer I got was no. No. Not right now. Long before Apple got into the game, I pitched the concept for an early iPod-like MP3 player to RealNetworks, Swatch, Palm. Everyone turned it down. No. No. Maybe we'll consider it next quarter. Maybe next year.

But when you're CEO, you're in charge. Sure, you're constrained by money or resources or your board, but for the first time there are no constraints on your ideas. You get to finally test out the things that other people told you couldn't be done. It's your opportunity to put your money where your mouth is.

That freedom is thrilling and empowering and utterly terrifying. There is nothing scarier than finally getting what you want and having to take responsibility for it, good or bad. And the tables begin to turn—as CEO you can't say "yes" to everything. You have to become the one who says "no." Freedom is a double-edged sword.

But it's still a sword. You can use it to cut through the bullshit, the hesitation, the red tape, the habituation. You can use it to create whatever you want. The right way. Your way.

You can change things.

That's why you start a company. That's why you become CEO.

Chapter

6.2

THE BOARD

Everyone needs a boss to be accountable to and coaches who can help them through difficult times—even a CEO. Especially a CEO. That's why businesses have a board of directors—typically just called "the board"—where members are directors of the company.

A board's primary responsibility is to hire and fire the CEO. That is the main way they can protect the company and the one and only job they have that really counts. Everything else comes down to giving good advice and respectful, no-bullshit feedback that hopefully steers the CEO in the right direction.

Ultimately it's the CEO who's responsible for running a company. But CEOs need to prove to the board that they're doing a good job, or risk being fired. That's why board meetings are so important, and why truly understanding your subject matter and doing a great deal of prep work beforehand is vital. The best CEOs always know the outcome of a board meeting before they walk through the door.

••

Bad CEOs come to board meetings and expect the board to help them make decisions.

Good CEOs walk in with a presentation of where the company was, where it is now, and where it's headed this quarter and in the years to come. They tell the board what's working but they're also transparent about what isn't and how they're addressing it. They present a fully formed plan that the board can question, object to, or try to modify. Things might get a little heated, a little bumpy, but in the end everyone walks out of the meeting understanding and accepting the CEO's vision and the company's path forward.

Then there are the great CEOs. With great CEOs the meeting is smooth as butter.

Watching Steve Jobs in an Apple board meeting was like watching a master conductor direct an orchestra. There was no confusion, no conflict. The board members already knew most of what he was going to say so they could just smile and nod. Occasionally someone would start a "what if" discussion and Steve would calmly let it meander for a couple of minutes, then say, "Let's talk about this outside the meeting, we still have lots to cover" and everyone would quietly settle down. Then in grand Steve style he'd whip out something fun and exciting to surprise the board—a new prototype or a never-before-seen demo. Everyone would walk out of the room happy and confident that Steve was leading Apple in the right direction.

Bill Campbell helped me understand how he did it. Bill would always say that if there was any potentially surprising or controversial topic, the CEO should go to every board member, one-on-one, to walk them through it before the meeting. That allowed them to ask questions, offer different perspectives, and then the CEO had time to take those thoughts back to the team and revise their thinking, presentation, and plan.

There should only be good surprises in a board meeting—We've exceeded our numbers! We're ahead of schedule! Check out this cool demo! Everything else should be a known quantity. It's best not to debate new discussion items in the boardroom—there's just never enough time to cover them in detail and get to a resolution. It always goes nowhere.

This is particularly true for public company boards. It's mostly because of their sheer size—they can have more than fifteen members, which makes meaningful debate more or less impossible—but also because of all the red tape and the laws surrounding them. Public boards are way more work for board members and execs alike and are infinitely more complicated than those at private companies. There might be up to ten additional committee meetings surrounding each board meeting, so the whole thing can take days.

(If a banker tries to convince you that going public is no big deal and won't radically change how you spend your time, do not listen to them. This is just the tip of the iceberg.)

Private company board meetings are shorter, typically quieter, more focused on the work and mentorship. They usually take two to four hours, sometimes five. It's less performative, less formal. Hopefully there are no committees in the early years at your startup, and only one or two (like an audit to oversee your financial statements) when you're in the growth stage.

The best thing about private boards is that you can keep them small—three to five board members is best. You can just have an investor, an insider, and an outsider with a specific expertise you really need.

But you also have to remember that even with a small board, the meeting's still not small. The room has twice as many people as you'd expect. In addition to the CEO and board members there's a lawyer, formal observers with some stake in the company, and informal attendees, like members of your exec team.

Before your first product ships, typically pre-revenue, meetings are pretty straightforward: you walk through anything pressing that needs board approval, then focus on your immediate progress in getting your product built. Where are we on the schedule? Are we spending money on budget? It comes down to what's going on internally and if you're on track to meet your goals.

After your product launch, and hopefully with revenue coming in, your board meetings will focus more on data and what's happening externally—what's the competition doing, what are customers asking for, how well are we attracting and retaining customers, what kinds of partnerships have you set up. And as always when you're presenting numbers, it becomes much more important to craft a narrative. You have to tell a story. [See also: Chapter 3.2: Why Storytelling.] Your board isn't in the business every day like you are—they can't immediately understand the nuances or what the numbers actually mean unless you give them context.

Being able to help the board grasp exactly what's going on is good for the CEO, too. The better you can explain something, the more you understand it. Teaching is the best test of your own knowledge. If you're struggling to explain what you're building and why, if you're presenting a report without really understanding it, if the board is asking you questions that you can't answer—then you have not internalized what's actually going on at your company.

That's when you might have a real problem.

It doesn't happen all that often, but sometimes a board does its most important and least pleasant job: they remove the CEO. Usually it's because the CEO is in the wrong—they're incapable or incompetent or pushing an agenda that will lead to ruin. Or sometimes a first-time founder has done a great job up until now but the company needs someone with different expertise and skills to take it to the next level.

But sometimes the problem isn't the CEO. It's the board.

The famous Tolstoy line "Happy families are all alike; every unhappy family is unhappy in its own way" applies to boards as well. Happy, functional, effective boards are all relatively small, full of experienced operators who have built companies before, think of themselves as mentors and coaches, and actually do the work—they help you recruit and get funding and expand your expertise, sharpen your business and product strategy, watch for land mines, and give it to you straight when you're about to step on one.

Bad boards come in all shapes and sizes and screw up in a million different ways. But they generally fall into three categories:

1. **Indifferent boards** occur when a majority of the board members are checked out. Sometimes an investor sits on a bunch of different boards and has a "you win some, you lose some" mentality—and has already put your company in the loss column. Sometimes board members are motivated for the wrong reasons—they want their payout and don't really care about the company or its mission. Sometimes they see obvious issues with the CEO but just don't want to do the work to remove them. Because it is work—paperwork and emotional fallout, then the search to replace the CEO, the interviews, the headaches, the internal transitions, the press, the cultural crises. They say, "It's not really that bad, is it?" and everyone suffers with the status quo because nobody feels like stepping up to fix it.
2. **Dictatorial boards** are the opposite—too invested, too controlling. They hold the reins so tightly that the CEO doesn't have the freedom to lead independently. Many times the board includes a previous founder (or two or three) who still wants control. So the CEO ends up behaving more like a COO—taking orders, fulfilling requests, keeping the trains running but not having much say in where they go.
3. **Inexperienced boards** are made up of people who don't know the business, don't know what a good board or CEO looks like, and are incapable of asking the CEO hard questions, never mind removing them. These boards are often too scared to act decisively. The investors worry that if they challenge the CEO, they won't get to invest in the next funding round or they'll get a reputation for firing founders and new startups won't want to work with them.

Typically companies with inexperienced boards are always running out of money. They never meet their quarterly goals and always blame "market problems" rather than the CEO or themselves. They don't know how to bring in new talent and new expertise and just smile and nod their way to collapse.

But even when a board isn't great—when it pushes too hard or not hard enough, when it makes the wrong call—it's a necessary part of any company's infrastructure. It needs to exist.

One of the most painful parts of the Google acquisition of Nest was losing our board. We had an amazing board at Nest—structured and informed, operational and active. We could go to the board, get firm agreement on a clear strategy and plan: yes, we're going to do this, I'll get back to you in a week with next steps.

When we were acquired, our beloved board was dissolved and replaced with . . . nothing. We were supposed to have a governing board of several Google execs, but our meetings were either perpetually rescheduled or barely attended. We'd propose a path forward and everyone would say, "Yeah, well, let's think about that a little bit more." The can would get kicked down the road to the next meeting that nobody went to and we'd be left sitting on our hands.

One might look at that and say, "So what's the problem? If the board doesn't give you guidance, then just go do it yourself. You're the CEO."

But that is not the solution. Even the most incredible CEOs in the world still need a board. Not the meetings, necessarily, but the advice of smart, invested, experienced people. Even big projects within a company should have a mini-board—a collection of helpful execs who can work to guide a project lead and step in if things go sideways.

I once saw an early-stage startup with a five-person board where the CEO controlled four seats. The CEO simply installed employees and friendly outsiders in all the available spots, and if anyone voted in a way the CEO disliked, they were removed. The one board member who knew better was completely helpless.

The CEO had complete freedom to follow their vision, to run things exactly as they wanted, to build the product of their dreams. Until they belittled their team, screamed at customers, and drove the business into the ground.

Countless millions were lost, and many people quit, but the most painful thing was the stupid waste of time and resources. It was so much needless strife.

Even the best CEO cannot stand alone, untouchable, unchallengeable, accountable to no one. Everyone needs to report to someone, even if it's a two-person board that you meet with for an hour every few months.

There always needs to be some kind of pressure-release valve. There always needs to be someone who can shake their head and give it to you straight.

And if you do it right, you should never be a victim of your board. As CEO, you help to shape it. Boards always change based on the CEO—the board under Steve Jobs was different from the board under Tim Cook. Boards complement a CEO's strengths and no two CEOs are alike.

So when you're choosing your board members, here's what types of people you should be considering:

1. **Seed crystals:** Just as you need seed crystals to grow your team, you want someone on the board who knows everyone, has done it before, and can suggest other amazing people to add to the board or to your company. [See also: Chapter 4.2: Are You Ready?: Seed crystals.] A seed crystal points out what your board is missing and tells you who to call, or just calls them for you. Our seed crystal for the Nest board was Randy Komisar—he's the one who

first suggested bringing in Bill Campbell. And he's the one we'd come to if we needed help with recruiting or landing a perfect candidate.

2. **A chairperson:** This isn't a must, but it can be helpful. A chairperson sets the agenda, leads the meetings, herds the cats. Sometimes the CEO is the chairperson, sometimes another board member is, sometimes there's no formal chairperson at all. I've seen all three versions work. But what worked best for the Nest board was having Randy Komisar as our unofficial chairman. Instead of me having to do all the 1:1s with board members myself, Randy would go talk to each one, pre-negotiate, and come up with the opinion of the group. He also interviewed execs for Nest and helped fill out the executive roster. A chairperson is the CEO's closest relationship on the board, a mentor and a partner. They help the CEO through issues they have with other board members, or they step in when the business gets hairy and the team gets scared. They'll attend employee meetings and give them the board's perspective on how the company is doing—they'll say, "The CEO's not going anywhere, she's doing a great job." Or "The board's not worried about the recent sales, and you shouldn't be, either. We can't wait to invest again." Or sometimes, "Yes this person has left, but the company will be okay. Here's the plan that the board supports."
3. **The right investors:** When you're picking investors, you're also picking one or two of them to be board members. So you don't want investors who think only in numbers and dollar signs and don't understand the slog and grind of creation. [See also: Chapter 4.3: Marrying for Money.] Find investors who are experienced in the work you do and empathetic to how difficult it is to get right. Find human beings who you'd love to have dinner with. If you have an interesting enough company, you can talk to your investors in advance and select the person the firm will put on your board.

Sometimes CEOs don't take the top-dollar investment deal, to ensure that they get a better board member.

4. **Operators:** These are people who have been in your position before and know the roller coaster of company building. When the investor board members start hammering you for not hitting your numbers, the operator board members can step in and explain the realities of the situation. They can talk about how nothing ever goes according to plan. Then they can help you forge a new plan with new techniques and new tools.
5. **Expertise:** Sometimes you'll need someone who deeply understands something very specific—patents, B2B sales, aluminum manufacturing, whatever—but they're too experienced or too entrenched in their current project to take a job at your company. So the only way to get them is through a spot on your board. When Apple was thinking about getting into retail for the first time, neither Steve Jobs nor anybody else on the board knew how to do it. So they brought on Mickey Drexler, the CEO of GAP. Mickey's the one who told them to get an airplane hangar and fully prototype a few different store designs, then walk through them in person like a real customer would before deciding which to take to the public. [See also: Chapter 3.1: Making the Intangible Tangible.]

The best board members are mentors first. They're able to offer sound, helpful advice at critical moments in your product's life or in yours. And they give as well as take—they enjoy the process of being on your board because they're learning something, too.

You just have to make sure they don't use those learnings against you.

When someone joins a board, they have a legal obligation to act in the best interests of the company they're serving. It's called a duty of care and a duty of loyalty. Usually people take this pledge seriously. But not always.

Sometimes people take advantage of their position. Sometimes they need to be removed. Sometimes it's dramatic.

But that's rare. Reshuffling a board is generally uncomfortable and tricky, but not impossibly so. You'll see it happen when you're a new CEO of an existing company and you inherit a board, or if you want to add a new expertise but not a new seat. The key is doing it in phases and with set time limits. First move a board member into an observer role for a couple of quarters, then move them out and the new person in. Doing it right takes time and patience.

And as always, even with the pressure, the flurry of meetings, the 1:1s and planning, you can't forget about your team. Board meetings are always moments of high stress for the entire business where everyone desperately wants to know what's going on and starts getting nervous about the outcome.

So don't make everyone wait and gossip and squirm. At Nest, most of the exec team knew exactly what was going on because they were in the boardroom with me, and we always showed a redacted version of the board presentation to the whole company as soon as possible after the meeting. Here's what we talked about, here's what I'm concerned about, this is what the board had questions about, here are actions we'll take.

It kept everyone in lockstep and quashed rumors. And if something changed, people were able to start working on those changes right away.

When you have a great board that you respect, meetings are a great, almost external heartbeat that focuses the entire company and forces you to organize your thoughts and timelines and story. [See also: Figure 3.5.1, in Chapter 3.5.]

It's worth it. But that doesn't make it any less work. For everyone.

That's why Jeff Bezos once told me never to join anyone else's board. "It's a waste of time," he said. "I'm only going to be on the board of my company and my philanthropy. That's it!"

I think of him every time I decline another board seat.

But I don't decline them all. My first instinct is always "No," but occasionally, rarely, that hard "No" turns into a "No. Unless . . ."

If you're trying to fill chairs and create the best board possible, remember that it's a two-way street. Most board members are experienced, busy, and highly sought after, so give them an incentive to join you. And I don't just mean stock. One of the best things about sitting on the board of a rising company is that you can get an early look into consumer behavior or new trends or disruptions. Anyone on the Apple board in the early 2000s, for example, got a preview of the iPhone and could plan ahead for what that product would do to their business.

Those kinds of insights are incredibly exciting for potential board members and the main reason people are always clamoring to get on the Apple board. Another reason is that they love Apple. They genuinely want to help the company succeed. They're willing to put in the time and effort because Apple is important to them.

Just keep in mind that public and private boards are very different. There's a lot more risk and work involved with being part of a public board, so you'll need a bigger reward to lure in the kinds of board members you'll need. Especially because most if not all of your early-stage board will probably resign when you go public. Public board members can get sued by shareholders. They have to go to countless committee meetings for audits, compensation, or governance. If things go sideways, they might get hammered in the press.

So the bar for taking a corporate public board seat versus an early-stage private board seat is very different.

Then again, any board seat comes with some level of cachet. It's good for the ego. Good for the pocketbook. But you never want that to be your main draw. Avoid celebrity board members, people who sit on a dozen or more boards or people who are only looking to get seats to pad their résumé. It's all too easy for them to check out. To become bored or indifferent. Or to put their interests above your company's.

You want board members who are truly, deeply excited by what you're making. Who can't wait to hear what you've been up to. Who aren't just there for the meetings but are with you day in and day out, helping you, finding opportunities for you to succeed. You want a board that loves your company. And that your company loves back.

Chapter

6.3

BUYING AND BEING BOUGHT

When two fully formed companies merge, their cultures need to be compatible. Just like any relationship, everything ultimately comes down to how well people get along, what their goals are, what their priorities are, and what drives them crazy. Fifty to 85 percent of all mergers fail due to cultural mismatches.

If a large company is acquiring a tiny team—a dozen people or fewer—cultural mismatch is much less of an issue. But even then, that tiny team should carefully assess how they'll be digested by the larger org and really take their time to understand the culture of the company they're about to join.

..

I don't regret selling Nest to Google. Neither does our executive team. We always revisit this question when our old team meets up. Our only regret is that we didn't get to finish what we started. But we made the decision to sell together, and we all stand by it today.

Given the data we had at the time, we'd do it again.

Especially since we were right. As predicted, once Nest brought the idea of the connected home to life, the giants of Apple, Amazon, and Samsung all wanted a piece. They built teams to compete with Google and Nest and created their own home products, platforms, and ecosystems. We dodged a bullet.

And Google was and remains an incredible company. It's filled with brilliant people at every level. It's changed the world many times over. Google's culture works for them—there's a reason a lot of people never leave the mother ship.

But that culture is enabled and driven by the fact that Google's search and advertising business pretty much prints cash. Even Googlers call it the "Money Tree." It's turned Google into a place of wild abundance where anyone can more or less do anything—or sometimes nothing at all. They've been so profitable for so long and have had so few existential business threats that they've never had to cut back or slim down, never had to be scrappy. They haven't had to really fight for anything in decades. Lucky them!

But at Nest, we were fighters. Our culture was born from the Apple way, a culture that survived multiple near-death experiences over its forty-plus years of existence. We were ready to fight for our mission and our place at the table, fight to keep our culture and our way of doing things.

Then, within hours of the acquisition, we had to fight for our customers. When they heard Google was buying Nest, customers panicked that they were going to be served ads on their thermostats. Newspapers shouted that Google would track your family, your pets, your schedule in their endless lust for data. So together, Google and Nest immediately put out a statement:

"Nest is being run independently from the rest of Google, with a separate management team, brand and culture. For example, Nest has a paid-for business model, while Google has generally had an ad-supported business model. We have nothing against ads—after all, Nest does lots of advertising. We just don't think ads are right for the Nest user experience."

It was the right thing to do for our customers. It was exactly the wrong thing to do for our relationship with Google.

On day one, with just a tweet-sized number (of very public) words, we innocently, naively alienated the company we had just joined. Many Googlers saw us as a band of fighters running at them, armed to the teeth and ready for war, already declaring independence, already rejecting Google's core business, and thought, Huh. What's up with them? Not very Googley.

The Google teams with whom we'd planned to integrate and codevelop technologies and products were reluctant to work with us. They kept asking their execs for more details to figure out if they really had to help us at the expense of their own projects. Why? Why? Why do we have to help a team that isn't Google? Over the subsequent months, every time we had to clarify yet again for customers that Nest was separate from Google, our internal reputation took another hit.

I should have remembered what it was like at Apple during the very first months when we started building the iPod. It just didn't occur to me—Nest was so much bigger and more established than my tiny iPod team, I thought this was a completely different situation. But it was exactly the same. Back then Apple's executive antibodies saw us coming to take their time and draw away their resources, so they tried to block our way and ignore our requests.

That's when Steve Jobs gave us air cover, dropped bombs on the teams who were slowing us down, forced the issue, yelled sometimes to make sure we got what we needed. Steve Jobs fighting for us was ultimately what allowed us to succeed.

But Google didn't have Steve Jobs. It had Larry Page and Sergey Brin—both brilliant and savvy entrepreneurs, but they didn't have Steve's fighting spirit, born from multiple near-career-death experiences.

At one point when the integrations we'd planned for completely stalled—when Googlers were literally not showing up to our meetings and ignoring our emails—Sundar Pichai told me that all the teams we were trying to work with were very busy. They didn't have extra cycles to dedicate to Nest. And no one at Google could simply dictate to them how to get things done—it was up to the teams to decide how to use their time.

As I stared at him, my eyes widened. I saw stars. It was as if I were in a car accident. Time slowed down. All I could think was, Oooooooooohhhhh, shhhhiiiiiiittttt.

I knew Google wasn't Apple and that a merger of this size would be bumpy. I knew we had different cultures, different philosophies, different leadership styles. But this is the moment that I realized we'd been speaking completely different languages.

When Larry told me during acquisition that Google would marshal the team and align their priorities with ours, he was 100 percent telling the truth. But what that looked like at Google was giving the team the skeleton of a plan and letting them fill in the rest as they went. Then they'd have a meeting every so often to ask how things were going.

But I had interpreted his words through an Apple lens. If Steve Jobs said he was going to marshal the team, that meant he was going to be there every step of the way—weekly, sometimes daily. He'd assemble everyone, tell them where to go, make sure they were marching together, and drag any stragglers back in place by sheer force of will.

Even though we were promised a full blitz, nobody was going to drop any bombs at Google. They didn't even know the meaning of the word.

The moment I realized that, I could see how we'd been misaligned from the start. We hadn't prepared for this. We hadn't planned for no managerial air cover. We hadn't planned for organ rejection.

Even while we'd meticulously planned out almost everything else.

In most acquisitions it takes two to eight weeks to draft a document with the necessary terms and get to an agreement.

With Nest it took four months.

And we didn't even discuss the sale price until ten weeks in.

Google Ventures, now known as GV, was an investor. They knew our financials and had always been extremely supportive, so I wasn't worried about the number. I was worried about which teams we'd work with, what technology we'd share, what products we'd build. Nest wasn't joining Google for the money—we were joining to accelerate our mission. So it was always mission first, money second.

Together with Google, we went through every single function—marketing, PR, HR, sales, every part of the company. We established where we could create synergies and where we couldn't, figured out which managers would be assigned to us, how we would do the hiring, which perks people would get, which salaries they could expect, which teams would be working together closely, and how those relationships would be established.

It took a lot of time. In fact I was starting to get a lot of eye rolls. "Really, Tony? You want to get into the details of this now?" Yes, yes, I do. It's important.

And it was—critically important and usually overlooked.

Most acquisitions are driven and overseen by bankers, and bankers only make the real money if the deal goes through, so they're motivated to move fast and get paid. They don't care about getting every detail of what happens to employees right. They don't really care about cultural fit. Not deeply.

Bankers are usually hired on both sides of the deal to work through all the transaction details, to help everyone understand or rationalize the price of the deal and comparable deals. They walk through the market, customer, and operational synergies.

But you can't pin down culture in a merger agreement. You can't write it up and have everyone sign on the dotted line. It's too squishy and sensitive, all about ineffable human relationships. And what bankers care about most is transactions, not relationships.

So most bankers don't want two companies to slowly feel each other out, get to know each other, date before marriage. They want them to meet and get engaged on the same night. They want them in a drive-through Elvis-themed chapel where everyone's a little liquored up so they don't ask too many questions. They want the deal wrapped up in thirty-six hours, before anyone has second thoughts, so they can pat each other on the back for a job well done and leave you standing in a blue ruffled tuxedo trying to figure out what's next. And if it doesn't work out in the long run, well, they did their jobs.

That's one of the reasons we didn't have a banker on our end of the Google acquisition deal. I knew bankers wouldn't care the way our team did. They'd be there to get a nice percentage of the deal for very little work compared to the years of blood, sweat, and tears from our team and investors.

Even so, the morning after the acquisition was announced, a banker showed up in the Nest lobby.

"I didn't see a bank representing you in the deal you announced yesterday."

"Yes—that was on purpose," I said.

"You know, your shareholders could sue you for that," he said.

I told him the deal was done and we didn't need a banker.

"Well, since you don't have a banker on this deal, could you just put our name on it?"

I raised an eyebrow, stared blankly at him, and walked away.

The banker was miffed. He couldn't believe I wouldn't do him the favor.

Most investment M&A bankers are not your friends. I've seen so many small startups, especially in Europe, call in a banker to help them raise money or sell their company. The bankers promise the moon and stars but rarely deliver.

You still may need a banker for various reasons, and of course there are a few good ones out there, but you can't let them control your merger or set your timelines.

Your job, whether you're buying or selling, is to figure out if your two companies' goals are aligned, if your missions nestle into each other, if your cultures make sense together. You have to consider the size of your companies. Can one be easily absorbed by the other? Is this a small team just starting out or a fully formed company with sales and marketing and HR and an ingrained way of working? If it's the latter, you have to understand what will happen to overlapping teams, what will change for employees, what will happen to your projects and your processes. That takes time.

And although time wasn't our issue at Nest, we did make a few major mistakes:

1. We put out a statement to customers without ever thinking about how it would affect our internal relationships.
2. I assumed that because our deal was really big—over $7 billion in total—there would be a standard of care and a fiduciary responsibility to make sure it was a success.
3. I listened to what Larry and Bill said about changing Google's culture rather than talking to employees about how firmly that culture was entrenched and what their expectations of Googleyness were.
4. I didn't talk to other companies that Google acquired before us.
5. We opened the door to Google employees who jumped from project to project and had no real interest in our mission or intention of staying when things got hard. They began to immediately dilute our culture and caused endless headaches complaining that we weren't Googley enough. [See also: Chapter 5.1: Hiring: But we were careful not to grow too fast.]

If I had talked to other VPs and directors at the company, I would have found out that we should have been much, much pickier about who we hired from the first wave of Googlers who reached out after acquisition. I only learned that I should have been wary six months later when friends at Google told me the unwritten rule: if you want to get great people off their teams, you have to fight for them. The ones who casually meander into your organization are just checking out the flavor of the month. And because Google is reluctant to fire people, many less-than-stellar performers just shift from team to team ad infinitum.

If I had spent more time with the leaders of previous acquisitions, like Motorola Mobility and Waze, I would have had a much clearer understanding of how Google digests the companies it buys. The majority of big Google acquisitions other than YouTube had been less than successful. As I soon found out, Google routinely jumped from shiny object to shiny object, and it didn't matter that Nest's price tag was in the billions. By the time we'd been consumed, they were already hungry again, moving on to the next meal. There was no time to make sure we were settling nicely in the belly of the beast, no interest in checking with us. We were just last night's dinner.

If I had talked to regular employees on the teams we wanted to integrate with, I would have found out what their priorities were and whether they were remotely interested in working with us. I would have understood better what it meant to be Googley and whether we had a chance of breaking through—and whether we could ever change what being Googley really meant.

Culture is incredibly sticky. I should have remembered that. Larry, with Bill Campbell's prodding, wanted Nest to come in and shift Google's entire way of thinking, to give it a burst of startup mojo. But culture doesn't work that way—you can't repaint an old factory and show the workers a training video and think you've made any kind of difference. You have to tear the whole thing down and rebuild it again.

Most people and companies need a near-death experience before they can really change.

You can't assume acquisition will mean acculturation. That's why Apple doesn't really buy companies with large teams. They only acquire specific teams or technologies, usually very early in their life cycle when they're pre-revenue. That way they can easily be absorbed and Apple never has to worry about culture. They can also skip the inevitable duplication of functions between existing teams like finance, legal, and sales, or the painful process of integrating one large team into another. With the notable exception of the Beats acquisition, Apple has been laser focused on filling small, specialized technology gaps in their evolving products rather than acquiring whole new lines of business.

All acquisitions come down to what you're trying to do when you're buying a company—do you want to buy a team? Technology? Patents? Product? Customer base? Business (that is, revenue)? A brand? Some other strategic assets?

When you're selling, the same questions apply. What are you looking for? Some hope to use a larger company's resources to accelerate their vision. Others are looking for financial gain. Then there are the companies who are having problems and are trying to sell off the business to someone who believes in it. Bill Campbell liked to say, "Great companies are bought, not sold." If you're being acquired, you want the buyer to be desperate to buy rather than you being a seller desperate to sell. If you're considering an acquisition, you want to be wary of anyone throwing themselves at you, pitching you too hard.

However, there's no manual for a good acquisition. There are a million things to watch out for, but they change with every company, every deal. Just don't ignore the hard stuff simply because it's hard. Don't forget to talk about culture just because nobody quite knows how to talk about culture.

Unfortunately, you can't truly know a culture until you're in it. It's like dating—when two people are interested in each other, they put their best foot forward, keep up appearances. Things get much more real when they move in together and get married. That's when you learn your wife leaves dishes in the sink to "soak" for a few days. It's when you realize your husband always forgets to clean up his toenail trimmings.

So the dating phase of any potential acquisition is crucial. You have to check the sink for dirty dishes. You have to spot the toenail on the dining table. Look at the reporting structure and the way they hire and fire employees. Dig into what perks everyone gets. Talk about management philosophy. Make concrete plans for exactly what's going to happen post-sale. Are you going to integrate or keep your cultures separate? What will you do about overlap? Where will this team go? Who will work on this product?

But always know that you won't be able to predict the future. Things will change—maybe in your favor, maybe not. And so, eventually, you just have to do it. Sign on the dotted line. Trust that it'll work out.

My advice is to always be cautiously optimistic. Trust, but verify.

Assume people have the best intentions, then make sure they're following through on them. And take the risk. Leap. Buy the company. Sell the company. Or do neither. Just follow your gut and don't be scared (or, rather, be scared but make the decision anyway).

If we hadn't sold, who knows what would have happened? Nest might have been successful on its own, or maybe we would have gone under as the big entrants inevitably showed up. Or maybe the other major players wouldn't have started working on their connected products and the entire ecosystem would have collapsed. Who knows? Can't run the experiment twice.

And Nest isn't dead. Far from it—it's alive and well. It's Google Nest now, fully integrated just like we'd always planned. And they're still making new products, creating new experiences, delivering on their version of our vision. It didn't turn out exactly like we wanted, but it was a tremendous learning experience and, damn, we got 70 percent there. Nest is still going, still building, and there's nothing I feel about that but happiness.

I bumped into Sundar Pichai, now CEO of Alphabet and Google, at a party a couple of years back. He said, "Tony, I wanted to let you know that we made sure to keep the Nest brand and the name. Nest will definitely be a part of our future strategy." I smiled wide and thanked him, touched that he went out of his way to tell me. Sundar is a class act and I'm grateful that he's watching over the team.

I have a lot to be grateful for.

I'm grateful that Sergey Brin pushed Google to invest early in Nest, grateful that both Larry and Sergey spurred the Google machine to buy us. I'm grateful that the other industry titans have come to focus on smart home technology, and that a hundred tiny upstarts are trying to displace them. In the end that's what will—eventually, circuitously—allow someone to reach our vision.

And also, I know that everything that happened after the acquisition wasn't personal. Not really. It was just business. Shit happens. No grudges. Life's too short.

I deeply, genuinely wish them well.

Chapter

6.4

FUCK MASSAGES

Beware of too many perks. Taking care of employees is 100 percent your responsibility. Distracting and coddling them is not. The cold war of ever-escalating perks between startups and modern big tech has convinced many companies that they need to serve three gourmet meals a day and offer free haircuts to attract employees. They do not. And they should not.

Keep in mind there's a difference between benefits and perks:

Benefit: Things like a 401(k), health insurance, dental insurance, employee savings plans, maternity and paternity leave—the things that really matter and can make a substantive impact on your employees' lives.

Perk: An occasional pleasant surprise that feels special, novel, and exciting. Free clothes, free food, parties, gifts. Perks can be completely free or subsidized by the company.

Getting benefits right is crucially important for your team and their families. You want to support the people you work with and make their lives better. Benefits allow your team and their families to stay healthy and happy and achieve their financial goals. This is where you should be spending your money.

Perks are a very different matter. In and of themselves, perks are not a bad thing. Surprising and delighting your team is wonderful and often necessary. But when perks are always free, appear constantly, and are treated like benefits, your business will suffer. An oversupply of perks hurts a company's bottom line and, contrary to popular belief, employee morale. Some people can become obsessed with what they can get rather than what they can do—believing perks to be a right, not a privilege. Then when times get tough or when the perks don't scale, they become outraged that their "rights" are being taken away.

And if the primary way you're attracting talent is through perks, then times will absolutely get tough.

••

A friend once proudly told me, "I bring my wife flowers every week."

He expected my admiration, I think. What romance! What generosity!

I said, "What?! I would never do that."

I bring my wife flowers from time to time, but it's always a surprise.

If you constantly give someone flowers, after a few weeks they won't be nearly as special. After a few months she'll barely give them a second thought. Every week she'll steadily lose interest.

Until the moment you stop.

You should absolutely do nice things for your employees. You should surely reward them for their hard work. But you have to remember how the human brain works. There's a psychology to entitlement.

If you want to give employees a perk, keep in mind two things:

1. When people pay for something, they value it. If something is free, it is literally worthless. So if employees get a perk all the time, then it should be subsidized, not free.
2. If something happens only rarely, it's special. If it happens all the time, the specialness evaporates. So if a perk is only received occasionally, it can be free. But you should make it very clear that this

is not going to be a regular occurrence and change up the perk so it's always a surprise.

There's a huge difference between giving people free food all the time, free food occasionally, and subsidizing food. There's a reason Apple provides subsidized meals rather than free ones. There's a reason you can get discounted products, but not free ones, when you work there. Steve Jobs almost never gave out free Apple products as gifts. He didn't want employees to devalue the very things they were working on. He believed if they are worthwhile and important, then you should treat them as such.

At Google, all employees used to get a holiday present of a free Google product every year. A phone or a laptop or Chromecast—something substantial. And every year, people moaned and complained—that's not what I wanted, that feels cheap, last year's was better. And then when they didn't get a present one year, there was outrage. How dare they not get us a gift! We always get gifts!

Free will screw you every time. Getting a really great deal on something creates a completely different mindset than expecting to get it for nothing.

Subsidizing perks rather than giving them away is obviously much better financially for your business, too. Companies that bubble-wrap their employees with tons of free perks are usually shortsighted and have no long-term strategy to sustain those perks, or they have an innately problematic core business and the perks are the cover. Facebook famously takes great care of its employees, but it also makes all its money selling customer data to advertisers. If Facebook changed its business model, their profitability would take an enormous hit and all those perks would disappear.

The trend of giving employees everything they could possibly want or need at the office originally started with Yahoo and Google. The idea came from a good, noble, honorable place—a desire to take care of people, an urge to make their company welcoming and fun. It was designed to make the office feel like college, better than college—a soft, comfortable place that you could settle into. And because Google has been making money hand over fist for so long (by selling its customers to advertisers, of course), the rest of the world thought this culture must be part of the reason. So the culture spread. Now the vast majority of startups in Silicon Valley offer gourmet meals, endlessly filled kegs, yoga classes, free massages.

But unless you have Google profit margins and revenue growth, you should not be giving Google perks.

Google shouldn't even be giving Google perks.

They've been trying to cut costs for years—they even started giving people smaller plates in their cafes to encourage them to take less food and cut down on waste. But once you set the precedent and shift people's expectations, it's almost impossible to claw your way back.

In the early days of Nest we had some snacks and drinks in the kitchen—mostly fruit. No packaged junk food. Why poison your talent? Once or twice a week we'd get the team tacos or sandwiches or something a little fancier for lunch. Once in a while someone would light the barbecue out back and people would stick around for dinner.

But with the Google acquisition came Google food. We built an enormous, beautiful cafe that served free breakfast, lunch, and dinner every day. Five or six different food stations offered different cuisines and menus and there were fresh pastries every morning. Cookies and cakes everywhere. Everyone thought it was really great. But it was really, really expensive.

After the skyrocketing costs of Alphabet, we tried to trim down some of the options at the cafe. Still plenty of amazing food but no more pho station. No more mini-muffins. There was immediate protest, a universal "What the hell? You can't take our mini-muffins!"

It was almost as bad as when we had to outlaw takeout containers after we realized a ton of people weren't staying late to work—they were hanging out until dinner, then shoveling a full meal into to-go boxes for their families and taking off.

The whole point of serving dinner was to reward employees who were working incredibly hard. But because it was free, people took advantage of it. It's free! It's ours! What's the big deal?

A couple of years before that, Taco Tuesday had been a treat. People were delighted when the fruit box got delivered. But now there was a new precedent.

And a new sense of entitlement.

I once saw a person stand up at TGIF, Google's weekly all-hands meeting—literally a meeting of tens of thousands of people—and complain that their preferred yogurt had disappeared from the micro-kitchens. These snack centers are required by Google to ensure that no employee ever has to walk farther than two hundred feet in search of food. This person felt it was their right, nay, their responsibility, to complain directly to the CEO, with all of Google as witness. About yogurt. Free yogurt. Why is the brand I like not within arm's reach? When is it coming back?

Just as any good and giving person can get taken advantage of, can be abused, so too can the good intentions of a company. Some people just take and take and take and believe it to be their right. And after a while the culture of the company evolves to accept that and even encourage it.

That's why I said, "Fuck massages."

When Google acquired Nest, I reluctantly approved the "all the time" free food and buses. They were part of working at Google, everyone already expected them, and they were genuinely helpful to our employees. I knew it would mean a cultural shift—I just hoped everyone would remember our scrappy roots. When we announced the Google acquisition to the team, I literally presented a slide that simply said, "Don't change." What got us there was exactly what we needed to continue. Just because we were changing investors didn't mean we should change our culture or what made us successful.

When Google gave us new, gorgeous, high-end office space after the acquisition, I thanked Larry Page. I said it was very beautiful. And I told him—and our team—that we didn't deserve it.

It felt wrong. We hadn't earned it yet. That building was meant for a profitable company that had already proven itself. It was meant for people who could relax and spend their time arguing about who was going to get the window seat, who'd get the best view. But that's not what Nest was about. We were focused on our mission, on staying late and solving problems and working hard and fighting through and over and around every obstacle in our path.

I wanted everyone to keep their focus on the things we were making, the vision we were trying to achieve. Not perks, not frills, not extras.

So there was no goddamn way that we were going to spend company money giving people free massages.

We needed that money—to build the business. To reach net margins. To make better products. To make sure our fundamentals were strong and sound so that we could keep employing all these people in the first place. And we needed it to help people have the life they wanted *outside* work. Instead of making the office so luxe that employees would never leave, we spent our money on meaningful benefits for them and their families—better health care, IVF, the stuff that really changes people's lives.

When we handed out perks, I wanted them to be purposeful in the same way. So we didn't try to trap people in the office—we rewarded employees by paying for dinner out with their families, or a weekend away. And we were happy to throw serious cash at stuff that genuinely improved people's experience, that brought them together and exposed them to new ideas and cultures and turned coworkers into friends. Anyone at Nest could join a club and request money to do something cool—barbecues for the whole company, a Holi celebration that painted half the parking lot, paper airplane battles that got more and more elaborate every week.

But as more Googlers joined our ranks and Nest employees started understanding what kinds of perks were typical at Google, there was a huge internal debate about what people were and weren't getting. Why did Googlers get massages? Why did they get more buses so they could come in late and leave after lunch? Why did they get 20 percent time (Google's famous promise to employees that they can devote a fifth of their time to other Google projects outside their regular jobs)? We want 20 percent time!

I said it wasn't happening. We needed 120 percent from everyone. We were still trying to build our platform and become a profitable business. Once we got there, we could talk about employees using Nest money to work on Google projects, get a free massage, and end their workday at 2:30 p.m. As you can imagine, my positions weren't popular with the new employees.

But there was no way I was going to let entitlement creep in when there was still so much left to do. I wasn't going to parcel out more perks just because Google employees were used to them.

The experience of being a Google employee is not normal. It's not reality. Clive Wilkinson, the architect of the massive, luxurious Googleplex, has even come to recognize it. He now calls his most famous work "fundamentally unhealthy." "Work-life balance cannot be achieved by spending all your life on a work campus. It's not real. It's not really engaging with the world in the way most people do," he said. [See also: Reading List: Architect behind Googleplex now says it's "dangerous" to work at such a posh office.]

It's the same problem that the very wealthy face—a gradual drifting upward, away from the regular problems of regular humans. Unless you stay grounded—take public transit, buy your own food, walk the streets, set up your own IT systems, understand the value of a dollar and how far it can take you in New York or Wisconsin or

Indonesia*—you start to forget the daily pains of the people you're supposed to be creating painkillers for. [See also: Chapter 4.1: How to Spot a Great Idea: The best ideas are painkillers, not vitamins.]

It's not just the customer that starts to go out of focus. As the number of perks increase, people's reason for being at their job can begin to blur as well. I've seen people who loved their jobs, found meaning and joy in making something, who worked hard but never felt like they were throwing away their time—until they fell headfirst into Google or Facebook or another corporate behemoth and completely lost their way. The more free stuff they saw other people get, the more they wanted. But getting those perks was only briefly satisfying—they lost their value over time. So they kept trying to get more and more. That became their focus. And making stuff, caring deeply about the work they were doing, creating something meaningful, really liking their job—it got lost by the wayside.

And it all started with the fucking massages.

To be clear, I fully support massages. I love massages. I get them all the time. Everyone should get massages. But at no point should your company culture be formed around the idea that massages are your due. At no point should you promise employees that they'll get massages forever. At no point should perks define your business or drag it down.

Perks are frosting. High-fructose corn syrup. And nobody will begrudge you a little sugar—everyone likes sweets from time to time. But stuffing your face full of them from morning to night isn't exactly a recipe for happiness. Just as dessert shouldn't come before dinner, perks shouldn't come before the mission you're there to achieve. The mission should fill and fuel your company. The perks should be a sprinkle of sugar on top.

---

* Take a look at www.gapminder.org/dollar-street to see how much people around the world make in a month and what their lives look like. It's an incredible resource to learn how different or similar we all can be.

Chapter

6.5

UNBECOMING CEO

A CEO is not a king or queen. It's not a lifetime appointment. At some point, you have to step down. Here's how you'll know it's time:

1. **The company or market has changed too much:** Some startup founders are not meant to be CEOs of larger companies. Some CEOs have the skills to manage one set of challenges and not another. If everything has changed so much that you have no idea how to manage it and the solutions you need to implement are completely out of your wheelhouse, it's probably time to go.

2. **You've turned into a babysitter CEO:** You've settled into maintenance mode rather than continually challenging and growing your company.

3. **You're being pushed to become a babysitter CEO:** Your board is demanding you stop taking big risks and just keep the trains running.

4. **You have a clear succession plan and the company's on an upswing:** If things are going great and you think one or two execs on the team are ready to move up, then it may be time to make some room for them. Always try to leave on a positive note and leave the company in good hands.

5. **You hate it:** This job is not for everyone. If you can't stand it, that doesn't mean you've failed. It just means you've discovered some-

thing useful about yourself and can now use that discovery to find a job you love.

••

We had to call a CEO's mom once.

My investment firm Future Shape had invested in his company—they had an incredible vision, a ton of potential—but the CEO was a first-time founder, completely unprepared for the job. He'd hear our feedback, fall on his sword, say it'll never happen again, and then of course it would. He never really listened and never really learned. After we had tried to coach him personally and professionally for more than eighteen months, things were only getting worse—he was demeaning staff in meetings, arguing in the hallways, even arguing with customers. Enough was enough. So the board fired the CEO.

But the CEO wouldn't leave.

We tried carrots. We tried sticks. He wouldn't budge, wouldn't listen to reason. Then he put a gun to our heads—he hired lawyers and began preparing to sue the board, the company, and his investors.

So we called his mom—the only person we thought he might listen to. We told her what would happen if we got sued: the board would fiercely countersue, the CEO's lies to investors would become public, and he'd probably never get funding for another startup again. He might not even get another job.

After almost a year of knock-down, drag-out fighting, that's what finally made him surrender. His mom.

But it was so bitter that we had to lock him out of the building and make sure he had zero further affiliation with the company. It was the only way to save an incredible team with a ton of potential to fulfill their mission.

At a different company, the same conversation took two minutes. We told the CEO that he shouldn't be CEO anymore. He sighed, then smiled. "Thank you," he said. "What a relief!"

Because a handful of founder CEOs have become famous and mind-bogglingly rich, there's this myth that the transition between starting a company and running it through all its phases, good and bad, is natural. Inevitable. If you create a startup, of course you're going to stick with it as it blossoms into a real company, then a corporation. Isn't that the whole point?

But a startup with five smart friends is a completely different beast than a company of 100, never mind 1,000. The job and responsibilities of an early founder and later-stage CEO are polar opposites.

Not every founder is cut out to be a CEO at every stage of a company.

Sometimes they don't know how a medium-sized company works, let alone a big one. They might not have the right mentors around them, might not know how to build a team or attract customers. And when all that falls on their heads, they typically revert to what they were good at when they were an individual contributor and abandon the real responsibilities of CEO, ignore the board's warnings, flounder, and implode. It's a hard but valuable lesson and many entrepreneurs learn from it and try again, usually with more success. I was one of them.

But that kind of experience is avoidable. You can notice when you're in a nosedive, you can look around, feel the wind in your hair. And you can do something about it—admit what's happening and step down.

But most CEOs who are on the brink of failure just shut their eyes and wait for the crash. There's often so much of their ego wrapped up in being a CEO, so much time and work. People spend their entire lives striving to lead a company. They make it the center of their self-worth and their identity. The prospect of letting go of that—just walking away—can be terrifying.

That's true when you're a first-time founder or when you've been leading companies for decades. Ego is a hell of a drug.

That's why some CEOs—even founders—just turn into barnacles. I can't tell you how many longtime CEOs I've seen clinging to the job even when their passion for it was gone. When they've slowly morphed from a parent CEO into a babysitter whose only interest is to protect what they've built, to maintain their position and the status quo. [See also: Chapter 6.1: Becoming CEO.]

These CEOs fool themselves into thinking it's okay that they don't feel the same intensity anymore, that they worked hard in the beginning and now they get to sit back and enjoy the spoils.

But that's not how it works.

It's your job as a CEO to constantly push your company forward—to come up with new ideas and projects to keep it fresh and alive. Then it's your job to work hard on those new projects, to be as passionate about them as you were about the original problem you came there to solve. In the meantime, other people on your team focus on your core business, optimizing the pieces that are already set up.

If you can't get excited about that, if you can't come up with new ideas or embrace the daring ones that your team dreams up, then that's a pretty good sign: you've become a babysitter. It's time to go.

There's no challenge to being a babysitter CEO. No joy. And worse—it's bad for the team. Bad for the company.

That fact isn't always obvious to everyone, though. Sometimes the board forces CEOs to act more like COOs. Just keep everything steady, they say. Everything is working, why take the risk? Don't spook the shareholders. We know best. Just follow orders.

That's what I faced at Google. And that's why I left.

Not just because Google was trying to sell Nest or because they wanted me to stop acting like a parent CEO, but as a warning to my team. I was under a gag order, so I couldn't tell everyone that something was seriously wrong. But I could show them.

They say "A captain should go down with the ship." I say bullshit. If the ship is obviously sinking, then the passengers will probably notice—at that point it's the job of the captain to stay aboard until everyone's safely in the lifeboats. However, if you're a CEO or high-level executive and you can see the water level rising before everyone else, then it's your responsibility to signal clear and present danger to your team. And there's no clearer sign that something ain't right than you walking out the door.

Sometimes the only warning flag you can wave is your resignation letter.

And sometimes it's even bigger than that. Bigger than you. Bigger than your team or your company. Sometimes the entire market changes; sometimes the priorities of the planet evolve. These are the moments when the company a CEO was running before doesn't make sense in the world anymore. Oil and gas CEOs are looking down the barrel of this moment right now. So are automakers. It's time for a new model.

And we need new blood.

Smart CEOs see the change coming, sooner or later—either for them personally or for the company or for the world. And they create a succession plan.

You never know when the rug will be pulled out from under you. Maybe your entire industry will change or you'll be bored by your job or you'll get hit by a bus. That's why you make a will. It's also why you hire other executives and possibly a COO who you'd be comfortable handing the company to eventually.

Even in an emergency, you want the transition to a new CEO to be as fluid and confident as possible.

But it shouldn't take an emergency to push you out. You shouldn't take your success as an invitation to stay forever. Don't look around and see the incredible team you've built and the company you've grown and think, Yes. This is it. This is success. I'm not moving from here.

That's not how it works, either.

That incredible team you've built needs room to rise, and you're currently sitting in the top spot. If they don't see any potential for moving up in their career, they'll start to leave and find different opportunities.

And the good times won't last forever. Upswings will inevitably turn into downswings. And you want to leave when things are going well, when you can hand the company over proudly to the next CEO, not throw it to them in a panic as you get cut loose by the board.

As we're writing this, Zhang Yiming, the founder and CEO of ByteDance, the creators of TikTok, announced that he's resigning. TikTok has never been more popular. Zhang is experiencing a high that few CEOs ever reach. But he can see a change coming. And in this case, it's internal. He just doesn't want the job. It doesn't suit him. "The truth is, I lack some of the skills that make an ideal manager," he said. "I'm more interested in analyzing organizational and market principles."

That's the kind of self-awareness and reasonableness that makes for a great leader. He seems to be making the right move, motivated by his gut, not his ego.

And now he has options. He can quit completely and maybe start a new company. He can move up and join the board and still have plenty of influence over how the company develops.

Or he can stay at the company and just take a different position. Part of the founder CEO myth is that once you're a CEO, there's no going back—that nobody wants to leave the job once they've got it. But people can bounce back and forth.

However, if a founder steps down from the CEO job but stays with the company, things can get messy.

If the founder isn't very careful, they can create all kinds of problems for the new CEO and the executives they leave behind. The same goes for cofounders—they have to be aware of the conflicts they can create in the company just by voicing an opinion. Founders have to watch how they're perceived, what meetings they go to, what language they use, what suggestions they make, and how clear it is that they are suggestions, not directives. They have to be extremely clear on their roles. Otherwise they can unknowingly—or intentionally—create factions in the company, some following the founder, some following the CEO, everyone getting upset, confused, pissed off.

I watched it happen at one company—the founder stepped down, helped to choose a successor. But then the founder stuck around, roaming the halls and giving everyone random bursts of feedback. Nobody was quite sure if it was proscriptive or just a suggestion, if they should jump to it or just take it as friendly advice. "The CEO can always be replaced, but not the founder. So I guess I should listen to the founder?"

The CEO was frustrated, the team utterly confused. So they settled on a new plan—the CEO would run things and the founder would step back, only communicating through the CEO. It worked well, everyone breathed a sigh of relief, things started looking up.

That lasted two weeks.

Then the founder walked into the team meeting again. Everyone's faces just dropped. "No, no not again!" They were completely deflated. For fourteen glorious days they'd known what they were doing, knew who to talk to, knew what the plan was. But the future was completely up in the air yet again. People started quitting. No one had the power to tell the founder, "Get out and stay out. We love you and your ideas, but you're making everything worse."

Founders need to be aware that they can easily undermine the work of the CEO and core team. Even if a founder chooses to only be a member of the board, they still have to be careful—they're not leading the team anymore. They become a coach, a mentor, an advisor. Just one voice out of many.

That's always hard. But it's much worse when you have to cut all ties. When your baby gets thrown to the wolves and the only thing you can do is walk away. That is agonizing.

When I was leaving Nest, I called an all-hands meeting of everyone in the company. And all these incredible people, hundreds of passionate, brilliant people who had built this company with me—from nothing, from a sketch, from a leaky garage with a squirrel problem—sat watching me in anticipation. I looked at them, I teared up. And I told them I was done.

Then I had to let Google do whatever they were going to do next.

That's the real twist of the knife when you leave, especially if it's contentious—the new regime usually chops your projects into pieces just to leave their mark and show that nothing of yours remains. They even take the pictures of the founders and early team off the walls. You have to know it's coming and still walk away.

And then you have to mourn.

When you're a founder, leaving your company can feel like a death.

You pour so much of your time, your energy, yourself into this business and then all of a sudden it's gone. A limb hacked off. A friend you've loved dearly and grown up with, gone forever.

Your new life seems bizarrely empty. Quiet. Your days and nights were utterly consumed before and now . . . nothing.

You'll feel terrible. Awful. But don't just immediately jump into a new job to distract yourself. And resist the urge to worry that your market value is declining every day you're not working. This feeling is usually driven by self-doubt, not the reality of the job market. The world will not judge you for taking some time off. There is a real shortage of talent—especially smart, dedicated talent in the CEO chair—so if you want to go there again, don't think you can't.

But you have to go through the time and mental exercises necessary to process the experience, recover, and then learn from it.

There's a half-life to everything.

In my experience it takes most people about a year and a half before they can start thinking about something new. There's a reason people in some cultures wear black for twelve months after a death. That's how long it takes to come to terms with this kind of loss.

The first three to six months will creep by as you get over the initial shock, the denial and possibly anger, the gnashing of teeth and tearing of hair when you see what they're doing to your baby. It will also take you that long to get through the list of crap you've been meaning to do but had been ignoring because of work. Only after you exhaust that list can you stop getting distracted by the past and start getting bored. This is a necessary step. You absolutely need to get bored before you can find new things to be inspired by.

It will take another six months to start reengaging with the world. To stop caring quite as much about what went wrong. To start learning new things, finding your curiosity again.

Then for the next six months you can begin to look at your life with fresh eyes. Get distracted. Get excited. Start thinking about what's next. And you don't have to get right back on the same racetrack that you jumped off the year before. Just because you were CEO once doesn't mean you need to be CEO again. You can always find or create new opportunities for yourself. You can always learn and grow and change.

Take the time you need to become the person you want to be. Just like you did at the start of your career and at every fork in the road along the way.

CONCLUSION: BEYOND YOURSELF

In the end, there are two things that matter: products and people.

What you build and who you build it with.

The things you make—the ideas you chase and the ideas that chase you—will ultimately define your career. And the people you chase them with may define your life.

It's incredibly special to create something together with a team. From nothing, from chaos, from a spark in someone's head, to a product, a business, a culture.

If everything aligns, if the timing is right, if you get incredibly lucky, you'll fight to create a product you believe in, that has so much of you and your team bottled inside it, and it will sell. It will spread. It won't just solve your customers' pain points; it will give them superpowers. If you make something truly disruptive, truly impactful, it will take on a life of its own. It will create new economies, new ways of interacting, new ways of living.

Even if your product doesn't change the whole world, even if it has a modest scope and a smaller audience, it can still change an industry. Do something different. Shift customer expectations. Set the standard higher. It can make a market, a whole ecosystem, better.

Your product, this thing you create together with your team, can eclipse your wildest expectations.

Or, then again, maybe it won't.

Maybe it will fail.

Maybe you'll have your own General Magic—an incredible vision, a beautiful idea, toppled by bad timing, immature technology, a fundamental misunderstanding of your customer.

Or maybe your product will thrive but your business will disintegrate. You'll work and work to create your own company, pour your life into the never-ending grinder of people problems and org design and meeting after meeting after meeting. And then you'll hand this gleaming jewel over to people who promise to love it and polish it and help it shine—and they'll let it slip through their fingers, fall to the dirt.

That happens sometimes.

Success is not a guarantee. No matter how great your team. How good your intentions. How wonderful your product. Sometimes it will all fall apart.

But even if your product is dead or your company is dead, what you've made still matters. It still counts. You'll walk away having created something you're proud of. Having tried. Having learned and grown. You'll still carry your idea and its as-yet-unfulfilled potential; you'll still hold on to the opportunity, the chance to try again.

And you'll hold on to the people.

Today I still work with friends I met at General Magic. At Philips. At Apple. At Nest.

The products have changed, the companies have changed, but the relationships haven't.

And now my life is all about relationships. Now my product is people.

After Nest I started an investment firm—Future Shape. We call ourselves "mentors with money." We invest our own cash into companies that we think will dramatically improve society, the environment, or human health. And then we give them what all VCs promise but rarely deliver—personal attention. Real help when they really need it, sometimes before they know they need it.

Although the people I mentor have taught me much more than I could possibly teach them. I've learned about so many different industries and businesses, about agriculture and aquaculture and material science, about mushroom leather and bikes and microplastics. With every team or founder I mentor, another world opens up.

This job is just as meaningful as anything I've ever made, any object I've manufactured. These are incredible people and incredible people are the heart of any and all innovation. They're going to change the world. Fix the world. Helping them, investing in them, mentoring them is probably some of the most important work I've ever done.

When I look back, I realize it's the most important work I've always done.

The best part of being an exec at Apple or a CEO at Nest was the opportunity to help people. That was always the most gratifying experience: I could help the team take care of their families. I could help someone if they got sick, or if their kids or parents got sick. And we built a community, a culture of quality and determination and innovation, where so many people flourished, where brilliant people could create and fail and learn and thrive together.

Often they came to Nest doing one thing, and left knowing that they could do a hundred.

All they needed was a push.

The thing holding most people back is themselves. They think they know what they can do and who they're supposed to be, and they don't explore beyond those boundaries.

That is, until someone comes along and pushes them—willingly or unwillingly, happily or unhappily—into doing something more. Into discovering a well of creativity or willpower or brilliance that they never realized they had.

It's a lot like pushing past the first version of a product. You dedicate every minute, every brain cell to creating V1. Exhausted, you edge it over the finish line. But even though it took all you had to make it, V1 is never good enough. You can see its huge potential. You can see how much more it can be. So you don't stop at the finish line—you keep pushing until you reach V2, V3, V4, V18. You keep uncovering more ways that this product can be great.

Humans are the same way. But so many of us get stuck at V1. Once we settle into ourselves, we lose sight of what we can become. But just as products are never finished, neither are humans. We're constantly changing. Constantly evolving.

So you push. As a leader, a CEO, a mentor—you push even when people resent you for it. Even when you worry that maybe you've pushed too far . . .

But there's always a reward on the other side.

It's worth it to do things well. It's worth it to try for greatness. It's worth it to help your team, to help people.

And one day you'll get an email from someone you worked with—two, three, maybe ten years ago. And they'll thank you. Thank you for pushing them. For helping them realize what they were capable of. They'll say they hated you for it then, resented every minute, couldn't believe how hard they had to work, how you made them start from scratch, how you wouldn't let up.

But eventually they realized that moment had been a turning point, a jumping-off point. It changed the trajectory of their entire career. The things you built together changed their life.

And that's how you'll know you've done something meaningful.

You've made something worth making.

ACKNOWLEDGMENTS

Writing this book was easier than I thought. And so, so much harder.

Harder than making the iPod. Or iPhone. Or Nest Learning Thermostat.

The easy part was figuring out what to write about. Almost every day an entrepreneur would ask me a question—about storytelling, about breakpoints, about growing a team or managing their board. We'd talk through their issue, I'd give them some advice, then I'd put it in the book.

So many of the topics were simple—it mostly felt like common sense. And I wondered if this book needed to exist. But then the next day, someone else would ask me the same question. A week later, it would happen again. And again and again. I've frankly been getting sick of hearing myself tell the same stories over and over, week after week, month after month.

It became clear to me why I was doing this. Common sense is common, but it isn't evenly distributed. You can't take the obvious approach to building a team if you've never built a team. You can't intuitively understand marketing if you've been an engineer all your life. If you're doing something new, if you're trying something for the first time, you have to earn your common sense. It's hard-fought wisdom that you have to stumble into through trial and error, through trying and failing, or—if you're lucky—through conversations with someone who's done it before. Many times you just need someone to confirm your gut feeling and give you the confidence to follow it.

That's why I wrote this book.

And it's why Bill Campbell never did.

Bill was a better coach and mentor than I'll ever be. People were always after him to write his advice down, but he always refused.

I think the reason was that being a mentor, a coach, comes down to trust—to a relationship between two people. To give good advice, Bill had to know you, your life, your family, your company, your fears and ambitions. He focused on helping one person in the moment they needed it most and tailored his advice to exactly what was going on in their lives.

But you can't do that in a book.

It was the thing I struggled with most while writing—not knowing my audience, not knowing what each reader was going through. There's so much to talk about—way, way too much. The first version of this book was seven hundred pages long. And even then, it all felt too surface level. I could never dig in as deep as I wanted. I could give general rules of thumb and tell stories about what worked for me, but they wouldn't work for everyone. And sometimes they'd be completely wrong.

But I decided to write what I knew. I looked back over everything I did, everything I learned over the last thirty-plus years, and opened up the curtains—showed how the sausage gets made. It's been hard, but also cathartic. It's helped me process a lot of what's happened over my career.

I've accepted that sometimes I'll be wrong. And sometimes I'll piss people off. But if you're not pissing someone off, then you're not doing anything worth doing. If you're not making mistakes, you're not learning.

Do, fail, learn.

And hopefully over the ten years that I've thought about writing this book, I failed enough and learned enough to figure out what was worth saying.

And who was worth thanking.

First, a sincere (I mean it) thank-you to all the assholes, bad bosses, terrible teammates, crappy company cultures, awful CEOs, incompetent board members, and incessant schoolyard bullies. I would have never realized what I didn't want to be without you. No matter how painful those lessons were, truly—thank you.

You helped motivate me to become someone who could do better. Who could write this book. And writing it wouldn't have been possible without the incredible efforts and trust of:

My wife and kids—thank you for always being there, being my inspiration, my support, and my mentors (and suffering through all the loud calls).

My cowriter in crime, Dina Lovinsky—it was a roller coaster of positive and not-so-positive emotions and life events to pull this together. But if it ain't scary it ain't fun, right?

My tireless team who helped on the book—Alfredo Botty, Lauren Elliott, Mark Fortier, Elise Houren, Joe Karczewski, Jason Kelley, Vicky Lu, Jonathon Lyons, Anton Oenning, Mike Quillinan, Anna Sorkina, Bridget Vinton, Matteo Vianello, Henry Vines and the Penguin team—who have had to suffer through all kinds of endless, inane requests and questions.

My editor, Hollis Heimbouch, and her HarperCollins team—for putting up with the crazy first-time authors and countless missed deadlines while we tried to achieve some sort of (naive) perfection.

My agent—Max Brockman and the Brockman team and especially John Brockman (for hounding me for over a decade to write any book).

All the encouragement, support, and great ideas from friends and readers—Cameron Adams, David Adjay, Cristiano Amon, Frederic Arnault, Hugo Barra, Juliet de Baubigny, Yves Behar, Scott Belsky, Tracy Beiers, Kate Brinks, Willson Cuaca, Marcelo Claure, Ben Clymer, Tony Conrad, Scott Cook, Daniel Ek, Pascal Gauthier, Malcolm Gladwell, Adam Grant, Hermann Hauser, Thomas Heatherwick, Joanna Hoffman, Ben Horowitz, Phil Hutcheon, Walter Isaacson, Andre Kabel, Susan Kare (designer of the famous walking lemon, among a million other things), Scott Keogh, Randy Komisar, Swamy Kotagiri, Toby Kraus, Hanneke Krekels, Jean de La Rochebrochard, Jim Lanzone, Sophie Le Guen, Jenny Lee, Jon Levy, Noam Lovinsky, Chip Lutton, Micky Malka, John Markoff,

Alexandre Mars, Mary Meeker, Xavier Niel, Ben Parker, Carl Pei, Ian Rogers, Ivy Ross, Steve Sarracino, Naren Shaam, Kunal Shah, Vineet Shahani, Simon Sinek, David and Alaina Sloo, Whitney Steele, Lisette Swart, Anthony Tan, Min-Liang Tan, Sebastian Thrun, Mariel van Tatenhove, Steve Vassallo, Maxime Veron, Gabe Whaley, Niklas Zennström, Andrew Zuckerman—your frank comments and advice helped so much to shape this book and give us the confidence to continue through the tough weeks.

The teams at General Magic, Apple iPod and iPhone, Nest, and our Future Shape family of entrepreneurs—without you this book could have never happened. I've learned so much from you and you truly helped keep me honest.

The friends and teammates we've lost along the way—Sioux Atkinson, Zarko Draganic, Phil Goldman, Allen "Skip" Haughay, Blake Krikorian, Leland Lew. Steve. Bill. I often think of you and wish we'd had more time.

And you, our readers. Thank you for your confidence in me and for buying this book. Not just because I worked hard on it, but because it's supporting something bigger. We printed this book using green practices so we'd have a minimal impact on the world, but we're using the proceeds from it to have a much larger one. Everything I make from this book will be invested into a Climate Fund managed by my investment and advisory firm, Future Shape.

Go to TonyFadell.com to learn more.

And thanks again. I hope this book helped you in some small way.

Onward,
Tony

PS: I'm not sure if I'm going to ever go through this again and write another book, but if you think I should dig deeper, or offer different advice, or write about something completely new, I'm listening. Email me at build@tonyfadell.com.

READING LIST

Here are some of the books and articles that have helped me, my friends, and mentors, in no particular order:

Give and Take: Why Helping Others Drives Our Success, Adam Grant

In Praise of Shadows, Jun'ichirō Tanizaki

The Monk and the Riddle, Randy Komisar

Why We Sleep: Unlocking the Power of Sleep and Dreams, Matthew Walker

The Messy Middle: Finding Your Way Through the Hardest and Most Crucial Part of Any Bold Venture, Scott Belsky

The Perfect Thing: How the iPod Shuffles Commerce, Culture, and Coolness, Steven Levy

Creative Confidence: Unleashing the Creative Potential Within Us All, David Kelley and Tom Kelley

Trillion Dollar Coach: The Leadership Playbook of Silicon Valley's Bill Campbell, Eric Schmidt, Jonathan Rosenberg, and Alan Eagle

The Hard Thing About Hard Things: Building a Business When There Are No Easy Answers, Ben Horowitz

Super Founders: What Data Reveals About Billion-Dollar Startups, Ali Tamaseb

Thinking, Fast and Slow, Daniel Kahneman

Noise: A Flaw in Human Judgment, Daniel Kahneman, Olivier Sibony, and Cass R. Sunstein

Beginners: The Joy and Transformative Power of Lifelong Learning, Tom Vanderbilt

Range: Why Generalists Triumph in a Specialized World, David Epstein

How to Decide: Simple Tools for Making Better Choices, Annie Duke

The No Asshole Rule: Building a Civilized Workplace and Surviving One That Isn't, Robert I. Sutton

A Curious Mind: The Secret to a Bigger Life, Brian Grazer

The Defining Decade: Why Your Twenties Matter and How to Make the Most of Them Now, Meg Jay

Work: A Deep History, from the Stone Age to the Age of Robots, James Suzman

Crisis Tales: Five Rules for Coping with Crises in Business, Politics, and Life, Lanny J. Davis

Crossing the Chasm: Marketing and Selling Disruptive Products to Mainstream Consumers, Geoffrey Moore

Entangled Life: How Fungi Make Our Worlds, Change Our Minds & Shape Our Futures, Merlin Sheldrake

Simple Sabotage Field Manual, U.S. Central Intelligence Agency, United States Office of Strategic Services, 1944 (https://www.gutenberg.org/ebooks/26184)

Read the Face: Face Reading for Success in Your Career, Relationships, and Health, Eric Standop

"Architect behind Googleplex now says it's 'dangerous' to work at such a posh office," Bobby Allyn, NPR, https://www.npr.org/2022/01/22/1073975824/architect-behind-googleplex-now-says-its-dangerous-to-work-at-such-a-posh-office

"Why and how do founding entrepreneurs bond with their ventures? Neural correlates of entrepreneurial and parental bonding," Tom Lahti, Marja-Liisa Halko, Necmi Karagozoglu, and Joakim Wincent. *Journal of Business Venturing* 34, no. 2 (2019): 368–88.

INDEX

SUSTAINABILITY INFORMATION

This book is as green as we could make it. It's important—to me, to the planet, to the next generation of humanity—that we move beyond the status quo. Ten percent recycled paper isn't going to cut it.

My goal was a fully compostable book made with 100 percent post-consumer recycled materials, no harmful chemicals, and materials and printing processes that had zero carbon footprint and used minimal natural resources. Unfortunately we couldn't even get close to my ambitions.

Nutrition Facts

Serving size 1 book (416 pages)

Jacket Paper and Printing	**Chain of Custody Certified, Not Recycled Material, Not Compostable**
Hardcover Paper Wrap	**Chain of Custody Certified, Not Recycled Material, Not Compostable**
Hardcover	**Recycled Materials, 10% Post-Consumer Waste**
Hardcover Ink	**Soy Ink, Compostable**
Interior Printed Paper Inks	**Soy Ink, Compostable**
Interior Pages Paper Stock	**100% Recycled, Recyclable, and Compostable**
End Sheet Paper Stock	**Not Chain of Custody Certified, Not Recycled Material, Compostable**
Glue for Binding	**Bio-Based and Compostable**
Printing	**UV Printing Presses Where Possible**
Print Partners	**Carbon Footprint Minimized**
Waste Programs	**Vendors Recycle All Paper Waste, Other Waste Not Recycled**

Our publishers worked with me to find the most innovative, cleanest processes and materials in the industry. But sufficiently green options often didn't exist, or we couldn't even find out what processes were used. Many parts of the bookmaking business are still opaque and need to be disrupted. This industry has a long way to go to become 100 percent green. Just like every other business in the world.

So if you have an idea or technology to innovate in fiber management, printing, binding, or recycling, I'm ready to hear about it. And fund it. Get in touch at tonyfadell.com.

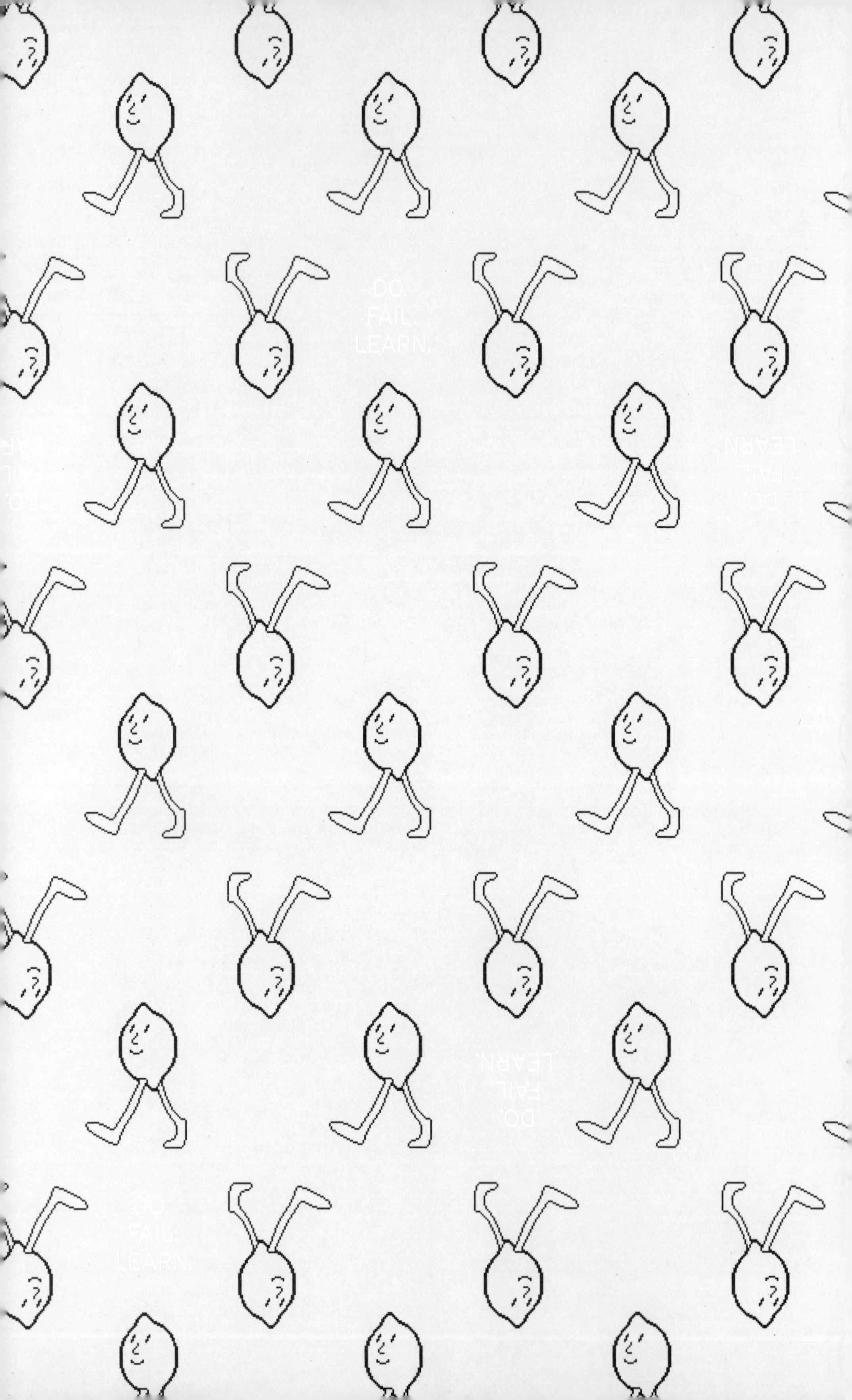
DO.
FAIL.
LEARN.
DO.
FAIL.
LEARN.